Science that changed our lives

Five scientific revolutions that changed
The way we live and understand the world

By Martin Gellender

ISBN 978-0-646-96731-8

Comments/feedback

The author invites comments and feedback, which can be sent by email to:
mgelled@bigpond.net.au

Other titles by the same author:

Science and the Big Issues of Our Time
  The science of how our world came to be and our choices for the future

National Library of Australia Cataloguing-in-Publication entry:

Creater:	Gellender, M.E., author

Title:	Science that changed our lives
	  Five scientific revolutions that changed
	  the way we live and understand the world/
	Martin Gellender

ISBN:	9780646967318 (paperback)
Subjects:	Social change Globalization - Social aspects Technological innovations - social aspects Discoveries in science - social aspects Science - 21st century

Science that changed our lives:

Five scientific revolutions that changed The way we live and understand the world

1. Flight: The technology that changed the 20th century

Introduction

My grandmother was born around 1890 and immigrated to the United States when she was 16 years old. I didn't know much about her, even though she was part of my childhood as – amazingly – she never learned to speak English! The area where she lived on the Lower East Side of New York would probably have looked like a film set of the Polish ghetto from where she started her life journey.

About the time that my grandmother arrived in New York, and when she was coming of age, the Wright Brothers developed the first powered aircraft. I suspect that my grandmother, and most people of the time, were completely unaware of this momentous development. Yet, probably more than any other technological development of the 20th century, aircraft became the iconic symbol of the modern age. Aircraft technology advanced rapidly, playing a crucial role in warfare, and then revolutionising long-distance transportation. I am nearly certain that my grandmother never flew in an airplane but, by the time she died, Boeing 747s were ferrying passengers around the world (as they continue to do today).

I grew up in a very different generation, and flying was part of it. In the 1950s and 1960s, flying on commercial jetliners became mainstream. Flying was new and exciting. It seemed amazing to me then that aircraft could overcome the bounds of gravity, carrying people above the clouds. It still amazes me. I was the first in my family (indeed, the first person I knew) to fly on a commercial jetliner. That was during my summer holiday from university in 1968, when I took a "stretch DC-8" on a charter flight to Europe and bicycled from Basel, Switzerland to Amsterdam. It was the first time I had been away from my parents, family and the world I knew. It opened my eyes to another world, and to the possibility of living in other places and other countries. It was an amazing adventure. For me, as an impressionable 20-year-old, it was nearly as exciting as flying to the moon.

My return flight from Europe was scheduled to depart early in the morning from Shipol Airport, outside Amsterdam. I had the clever idea to ride my bicycle to the airport on the previous afternoon and spend the night lying across some seats in the airport terminal. It seemed a good plan, but everyone else had the same idea! The airport terminal was like a refugee camp, packed with hundreds of passengers. I couldn't find an empty seat to sit, let alone several to lie across. Eventually, I did find a place to lie down, and crashed asleep. I awoke to the announcement of the final call for my flight. I opened my eyes and was amazed to see that the terminal was completely deserted! Not one person in sight. I ran to the empty departure area, where an attractive blond hostess was standing at the desk. As she printed my boarding pass, another staff-member approached and informed her that the Russian army had just occupied Czechoslovakia. I knew what that meant: I remembered being terrified watching television news coverage of Russian tanks on the streets of Budapest during the Hungarian uprising of 1956, when I was eight years old. I took my boarding pass and walked onto the aerobridge towards the plane. The door closed behind me. I was glad to be heading home.

Unlike earlier modes of travel, flying on modern jets connects distant parts of the Earth - with different seasons, climates and cultures - within an incredibly short period of time. This was brought home when I first flew to Australia with my new wife on our honeymoon in December 1976. I would be meeting my wife's family and seeing our future home for the first time. At that time, such long flights were broken into segments. We left Toronto Airport on a bitterly cold and bleak winter day, arrived in Vancouver in darkness and rain, and took off before dawn. Our plane headed west over the Pacific, following the darkness as the world turned. The cabin lights were dimmed, the passengers pulled down the window blinds, and I managed to crash asleep (I must have had the knack to do that). I was awoken by an announcement of our final approach into Honolulu. I opened my eyes, moved my seat upright and flipped up the window shade . . . and was nearly blown out of my seat by an explosion of sunlight! As my eyes adjusted to the brilliant blue sky, I could see palm trees waving in the tropical breeze. It was clear that I was arriving in a very different world from the one I had just left.

Like many in my generation, I have always been fascinated by aviation. As a scientist, I wanted to understand the physical principles that explain how airplanes fly. This was not straightforward. Most of the theory of flight was developed at the very beginning of the 20th century, when aircraft were small flimsy craft that flew as fast as a car on a suburban street. The theory made no intuitive sense to me, and seemed inconsistent with basic physics I learned in school. It assumed that air acts like an incompressible fluid which (for reasons I won't explain here) is reasonable and applicable at very low speeds, but becomes increasingly dubious and irrelevant at the speeds of modern jetliners. The explanation of how airplanes fly that I (and most of my generation) was told at school, and which has become accepted on faith, is clearly wrong or, at least, misleading. I suspect that, as aircraft became faster and heavier, aircraft designers forgot about the theory, and were guided by practical experience, wind tunnel testing and (later) computer simulations.

Over the years, I have read explanations that dissented from the existing dogma, and I have tied these together and "filled in the blanks" with my own ideas. This is the explanation that I present on the following pages, and commend to you. It allows a relatively simple and intuitive understanding of flight that is fully consistent with the principles of physics and with the known behaviour of actual aircraft.

Today, flying has become so common that it holds no mystique for younger generations. For many, it is routine and tedious (and I must confess, sitting in one seat for 13 hours on a long-haul flight is quite tedious). My son-in-law "commutes" from his home in Brisbane to his job in Kenya on an eight-week "fly-in, fly out" cycle. Each year, he flies about 200,000 kilometres, half the distance from the Earth to the moon. This was unthinkable a few decades ago, but has become "the new normal" for many professionals. As routine and tedious (and disruptive of family life) as this might be, long-distance flying has become integral to their livelihood and way of life.

2. Winged aircraft

When we were in primary school, most of us were told that airplanes fly because of the curvature of the wing (the arc along the front-to-back midline of the wing, or its "camber"). This curvature, my teacher said, causes air to fly faster over the top of the wing than the bottom, and this causes a pressure difference on the wing. As I later found out, the wings of many aircraft are cambered, but some wings are symmetrical – they have no camber at all. Somehow, planes with symmetrical wings fly just fine. Furthermore, many stunt planes and military fighter jets can fly upside-down just as well as they fly rightside-up. So clearly, the air doesn't "care" that much about whether the top of the wing is more curved than the bottom.

To understand how airplanes really fly, we should not be too pre-occupied with the shape of the wing. Rather, we should focus on the *air* as a wing flies through it. The purpose of an airplane wing is to accelerate the surrounding air downwards (just like the rotor of a helicopter), and in this way, generate an upwards lift force.

We have all seen videos of firefighters holding a high-pressure water hose, trying to extinguish a blaze. Sometimes it takes two or three burley firemen to resist the backwards force on the hose produced as large volumes of water shoot out at high speed. Rapidly-flowing water generates a "recoil force" in the same way that a marksman experiences a recoil "kick" when he fires a gun. In the case of firing a gun, a very high force is exerted on the bullet, pushing it down the barrel, for a tiny fraction of a second. Exactly the same force acts on the gun in the opposite direction, pushing it backwards, for exactly the same period of time. The firemen are not shooting a single projectile, but rather, a continuous stream of fluid, but the principle is the same. The force acting on each glob of water, accelerating it out of the hose, is exactly the same as the "recoil" force pushing the hose backwards. In fact, the backwards force is equal to the rate at which "momentum" is imparted to the water – that is, the mass per second of water multiplied by its velocity.

In theory, to make an airplane fly, we could simply shoot a continuous stream of water downwards. In fact, I have played with toy "water rockets" that work in exactly this way. But this would not be very practical for an aircraft because it requires a lot of energy to accelerate the water. That's because the downwards momentum of a flowing stream (which is what we want to produce an upwards lift force) varies with the *velocity* imparted to the stream, while the kinetic energy needed to accelerate the fluid varies with its *velocity squared*. So, to reduce the energy required to produce an upwards lift force, *we want to accelerate the largest possible mass of fluid downwards at the lowest possible speed*. An aircraft wing does this by accelerating large volumes of air downwards.

Imagine that you are a molecule in the air, just drifting around in the atmosphere minding your own business, when the wing of a Boeing 747 suddenly approaches at velocity **v**, say 250 metres/second. The front "leading edge" of the wing splits the airstream. You might find yourself passing over the top of the wing, while a neighbouring molecule below you (let's call him Bill) passes beneath the wing.

As it turns out, if you are close to the wing, your path follows the shape of the wing surface. If the wing slopes downwards (that is, has a positive "angle of attack" **θ**), you acquire a downwards velocity to follow the wing surface.

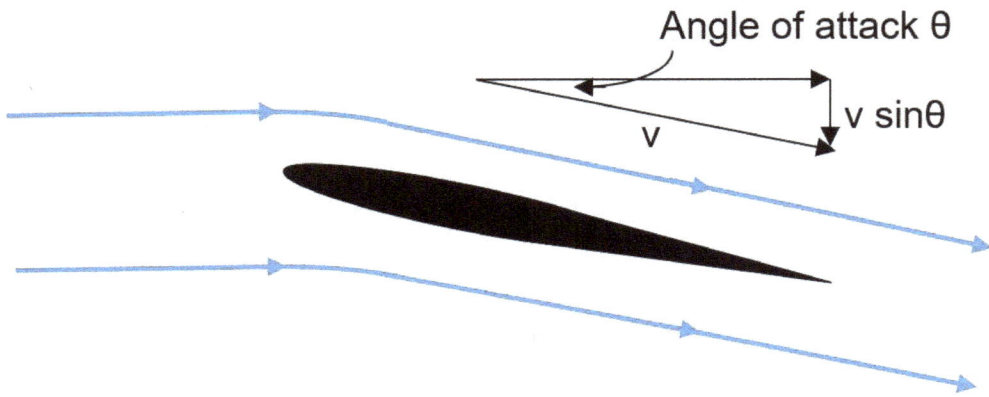

Angle of attack θ

v

v sinθ

Even after the aircraft wing flies past and recedes into the distance, life does not return to exactly the way it was before. You have been left with a downwards velocity, **v sinθ**, which you acquired as the wing passed beneath you. You peer down at your neighbouring

molecule Bill, who passed beneath the wing, and note that he too has been left with a residual downwards velocity.

The passage of a wing imparts downwards velocity, and downwards momentum, to the surrounding air. Air has mass, so a force is required to accelerate it downwards. An equal and opposite upwards force acts on the airplane wing. The upwards force on the wing, which is called "lift", supports the weight of the airplane. When an aircraft is cruising at constant altitude, the lift force is equal to the weight of the aircraft.

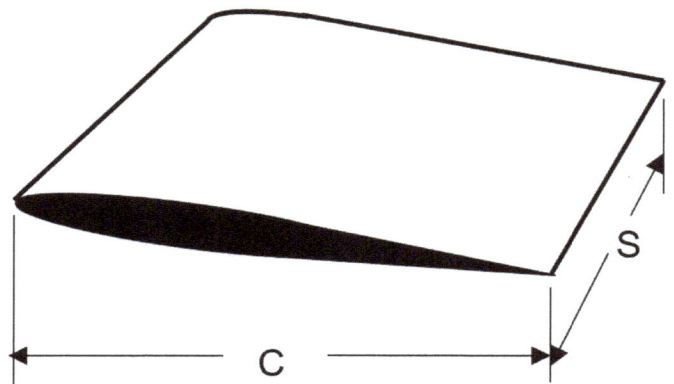

S

C

Let's see how much lift force is developed by a wing which has a "chord" length **C** from the leading edge to the trailing edge, and a wingspan **S**. Normally, the wingspan is much longer than the chord length.

I've mentioned that the air follows the shape of the wing, but the ability of the wing to deflect the airstream gets less as we travel out above and below the wing. We can imagine that just above the wing, the airstream will be exactly parallel to the wing surface, but as we move

away from the wing surface, the downwards velocity imparted to the air gets progressively less.

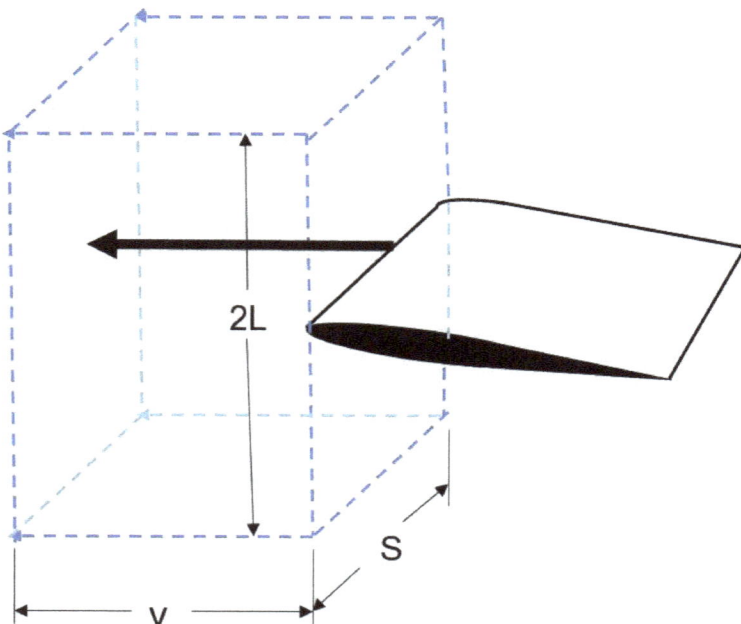

2L

S

v

The effect of the wing in deflecting the air extends over an "effective" distance that is roughly equal to the chord length **C** of the wing. So, the volume of air that is deflected by the wing lies in a rectangular volume[Note 1]. One side of the rectangle is the vertical distance extending distance **C** above and below the wing (a total distance of **2C**). Another side is the wingspan **S**. The rectangle extends forward at the velocity **v** of the aircraft, so each second, the rectangular "zone of influence" of the wing increases by distance **v**.

The total volume of air that is deflected each second is $(2C)(S)(v) = 2CSv$

The mass of air that is disturbed each second is given by multiplying the volume of deflected air by its density ρ_{AIR}.

Mass of air deflected each second = $(2C)(S)(v) = 2CSv$

You might be surprised at how much air is deflected by the wings of a large jetliner at cruise speed. A Boeing 747-400 has a wingspan **S** of 64 metres and an average wing chord **C** of 8.7 metres. It cruises at a speed of 250 metres/second (960 km/hour), with an air density of about 0.4 kilograms/m³ at cruising altitude. Each second, the wings deflect about 280,000 cubic metres of air – with a mass of 112 tonnes.

This mass of air is deflected downwards at velocity **v sin θ**. Normally, the "angle of attack" θ is relatively small (3-5 degrees). For readers who have studied trigonometry, you may recall that the value of **sin θ** is equal to angle θ, provided that the angle is small (and is expressed in units of radians). So, the downwards velocity imparted to the air is simply **vθ**.

The force required to deflect the air, which is also the upwards lift force applied to the wing, is equal to the rate at which momentum is imparted to the air – which is the mass of the air multiplied by its change of velocity.

$$\text{Lift force, } F_{LIFT} = \underbrace{(2\,C\,S\,v\,\rho_{AIR}\,)}_{\substack{\text{Mass} \\ \text{of air}}}\underbrace{(\,v\theta\,)}_{\substack{\text{Velocity} \\ \text{imparted}}}$$

Let's combine terms with velocity v. Notice, also, that (**CS**) is equal to the area of the wing **A**$_{WING}$. This gives:

Equation (1)	Lift force, $F_{LIFT} = 2\,\rho_{AIR}\,A_{WING}\,v^2\,\theta$

This is an amazing result! It explains a lot about how airplanes fly, and their actual flight characteristics. Equation (1) should seem intuitively reasonable. If the density of the air ρ_{AIR} increases, a greater mass of air would be deflected downwards, so we would expect the lift force to increase. The lift also increases with the **square** of the aircraft velocity **v**. If a plane flies twice as fast, twice as much air is deflected, and it is deflected downwards at twice the velocity.

Note also that the lift force varies directly in proportion with the angle of attack **θ**. This is exactly what is observed when the lift force is measured by placing a wing in a wind tunnel.

At constant velocity, the lift force increases linearly as the angle of attack is increased. But once the angle of attack exceeds a critical point, the lift tapers off, reaches a maximum, and then declines rapidly.

The lift reaches a maximum value when the angle of attack reaches the "stall angle" (usually at about 25-30 degrees), and then decreases rapidly. At such a steep angle of attack, the airstream no longer follows the surface of the wing. "Flow separation" occurs. Rather than smoothly deflecting the air downwards, the wing generates turbulence and pushes the air forwards (a bit like a snowplow). In this situation, the wing "stalls". This can be catastrophic, and normally pilots try to avoid stalling at all costs (although stunt pilots sometimes deliberately stall their aircraft in manoeuvres that they have carefully trained for and rehearsed).

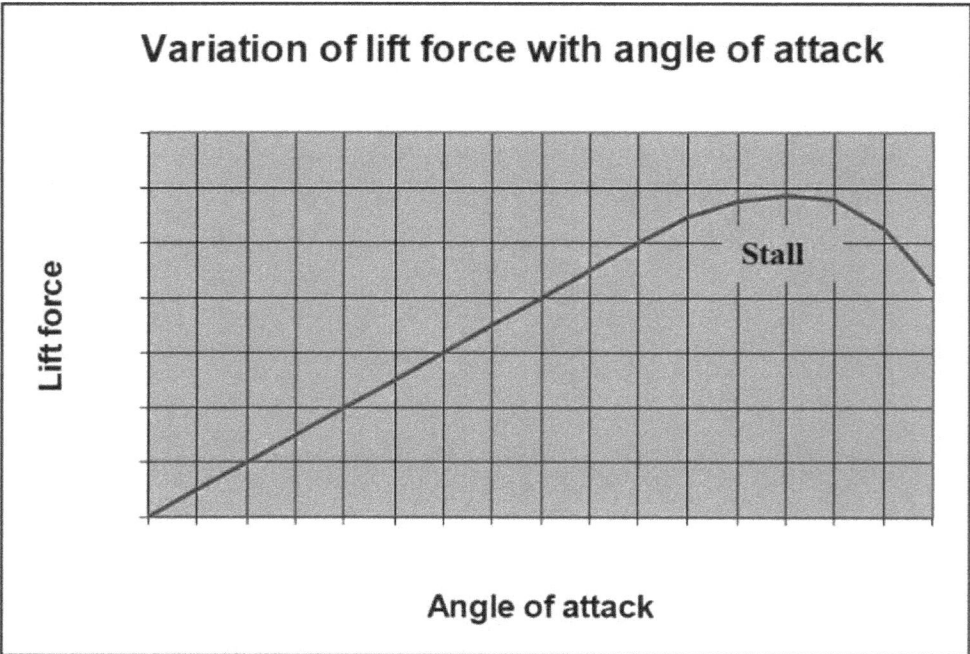

Variation of lift force with angle of attack

Lift force

Stall

Angle of attack

Normally, when a commercial jetliner is cruising at high subsonic speed (about 900 kilometres/hour), an angle of attack of a few degrees provides sufficient lift. Of course, it would be highly problematic to try to land a jet at such high speed. Generally, we want the landing speed of an aircraft to be as slow as possible, but this presents a dilemma.

As a jetliner approaches its destination, the pilot reduces the altitude and speed. Greater density of the air near ground level makes it easier to generate sufficient lift to support the weight of the plane. Near sea level, the density of the air is about three times that at cruising altitude, which would mean that the angle of attack could be reduced to a third – if the speed remained the same. But, the speed of the plane is also reducing on its landing approach. *The aircraft slows to about one-quarter of its cruising speed, which reduces the lift by sixteen times*. With the density of air increased by three times, and the velocity reduced to one-quarter, the angle of attack must be 16/3, or about 5, times greater than at cruising conditions.

To achieve the lowest possible landing speeds, without risking a stall, modern jetliners deploy flaps from the rear of the wing as they approach their destination. These flaps increase the area of the wing, and also increase the curvature of the wing so the plane can achieve a high angle of attack without the plane pitching nose-up.

We can re-arrange Equation (1) to relate the cruising speed of an aircraft, or any flying object (bird, bat or insect) to the "wing loading" (the weight W per wing area A_{WING}). What we get is:

$$\text{"Wing loading"}, \quad \frac{W}{A_{WING}} = 2\, \rho_{AIR}\, v^2\, \theta$$

If we look at the huge range of flying creatures and objects, they vary enormously in size and weight – from beetles weighing about one gram to a jumbo jets weighing 300 tonnes – a factor of 300 million! Yet, all of these hardly vary at all in their angle of attack while cruising, and the density of air only varies by about a factor of four (for beetles flying at sea level to commercial jets flying at cruising altitude). This means that, for the enormous range of flying objects – from insects to A380s - the "wing loading" varies with the square of the velocity. A Boeing 747 travels more than 300 times faster than a beetle, so the weight carried by each square centimetre of wing is about 100,000 times greater. This relationship is clearly evident from various graphs in the book "The simple science of flight: from insects to jumbo jets" by Hendrik Tanneker[Reference 1].

The same book also shows how a wide range of birds (from a one-kilogram common tern to an 80 kilogram wandering albatross) fly at cruising speeds that can be predicted from their weight. The author argues that birds have roughly the same proportion of body size and wing area (presumably because the optimal shape for flying was maintained by evolution) and the same density of body tissues. Most birds fly at relatively low altitude, and thus, experience the same air density. Presumably, all birds cruise with roughly the same angle of attack. Consequently, if a bird is twice as large (in length, width and height), it has eight times the mass, and four times the wing area, and Equation (1) indicates that its cruising speed would be greater by the square root of 2, or 1.4 times. Indeed, for a huge span of bird species, their "cruising speed" varies with the square root of their size, and with their weight to the one-sixth power. This means that an 80-kilogram albatross will fly at twice the speed of a one-kilogram tern.

Notes
1. The volume of deflected air is only rectangular if we neglect the region near the wingtips. However, since the wingspan is generally much longer than the chord length, effects occurring at the wingtips can be neglected, greatly simplifying the calculations.

References
1. Book available by request at State Library of Queensland. Call No. 629 1323.1996.

3. Power required for winged flight

Flying tends to be highly energy-intensive, for aircraft and for birds. Minimising the energy consumption for flight is a major challenge. For example, to transport one kilogram of cargo by air freight requires about eight times as much energy as to send it by truck, or fifty times as much energy to ship it by rail, or two hundred times as much energy to send it across the ocean on a large container ship.

Birds that migrate for long distances eat large amounts of food to build up reserves of body fat as "fuel" for the journey. Fat stores used as fuel on long journeys by migratory birds may account for 40% of their body weight on take-off, about the same fraction as the fuel consumed by jetliners on international routes.

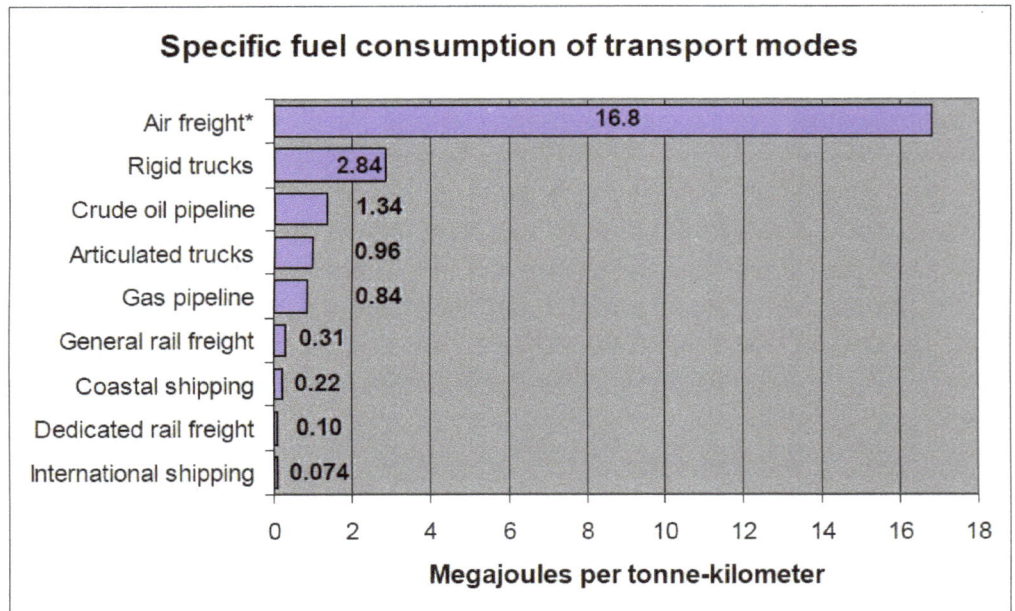

Specific fuel consumption of transport modes

Mode	Megajoules per tonne-kilometer
Air freight*	16.8
Rigid trucks	2.84
Crude oil pipeline	1.34
Articulated trucks	0.96
Gas pipeline	0.84
General rail freight	0.31
Coastal shipping	0.22
Dedicated rail freight	0.10
International shipping	0.074

* Based on 2.53 megajoules per passenger-kilometer, and attributing 0.15 tonnes per passenger (for passenger, seat, luggage, catering and toilet facilities. Data Source: Australian Transport Facts 2004.

Previously, we discussed how airplanes produce lift to fly. We'll now take this further to consider the energy required to fly. This is an issue of fundamental importance to the air transport industry, which transports growing number of people and increasing amounts of goods across the world, and now accounts for a significant fraction of the fossil fuel energy used by humanity.

In the previous chapter, I explained that an airplane wing develops lift force by deflecting air downwards. The lift force is generated by transferring momentum to the air, so the greater the mass of air that is pushed downwards, and the greater the downwards velocity transferred to the air, the greater is the lift force.

I explained earlier that the influence of the wing (in deflecting the air) extends above and below the wing for a limited distance. As it turns out, this distance is roughly equal to the chord length of the wing (from the front leading edge to the back trailing edge). Consequently, the volume of air that is deflected downwards is equal to the area of the wing A_{wing} times the forwards velocity v of the airplane. Multiplying this volume by the density of air ρ_{air} gives the mass of air deflected each second, which is:

$$\text{Mass of air deflected each second} = 2\,\rho_{air}\,A_{wing}\,v$$

Also, as described previously, the air follows the surface of the wing, which is inclined at the "angle of attack" θ. So, the downwards velocity imparted to the airstream is $v\theta$.

We derived the lift force by multiplying the mass of air deflected each second by the downwards velocity, to get the result:

Lift Force = $2 \rho_{air} A_{wing} v^2 \theta$

Bear in mind that the lift force holds up the weight of the plane, and during normal flight, the lift force is equal to the weight **W** of the aircraft.

We can re-arrange this equation to derive the angle of attack that is required for a plane to fly at a given velocity **v**.

Equation (1) Angle of attack, $\theta = \dfrac{W}{2\, \rho_{air}\, A_{wing}\, v^2}$

Now, this raises a fundamental dilemma. To produce lift, the wing must push the air downwards. But, ***pushing the air downwards imparts kinetic energy to the air***, and we (or rather, the aircraft engines) must do work to provide this energy. The process of producing lift causes a drag force pulling backwards on the wing, and a power input is needed to overcome this drag force.

Let's calculate how much kinetic energy is imparted to the air by a wing. Remember that the kinetic energy of any moving body is one-half its mass multiplied by its velocity squared. So, the downwards kinetic energy imparted each second to the air is simply one-half the mass of the deflected air ($2\rho_{air} A_{wing} v$) multiplied by its downwards velocity $v\theta$ squared. The result is:

Kinetic energy imparted to the air per second $= \frac{1}{2} (2\, \rho_{air} A_{wing}\, v)(v\, \theta)^2$

$= \rho_{air} A_{wing}\, v^3\, \theta^2$

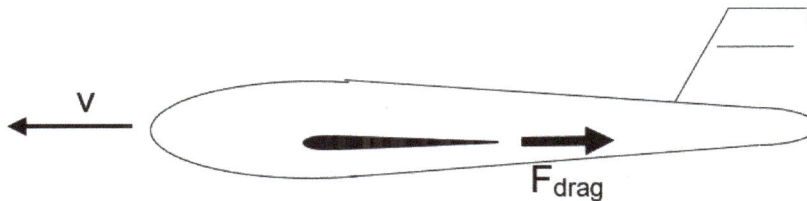

F_{drag}

This kinetic energy is provided by the aircraft engines pushing against a drag force F_{drag}. The drag force acts over the distance that the plane flies. So, the work that the engines must produce each second is equal to the drag force times the aircraft velocity.

$\underbrace{F_{drag}\, v}_{\substack{\text{Work done} \\ \text{by engines}}} = \underbrace{\rho_{air} A_{wing}\, v^3\, \theta^2}_{\substack{\text{Kinetic energy} \\ \text{Imparted to air}}}$

Cancelling out the **v** which appears on both sides of the equation gives:

$F_{drag} = \rho_{air} A_{wing}\, v^2\, \theta^2$

We have already derived Equation (1) for the angle of attack θ, so let's substitute this into the drag equation. This gives:

$$\text{Equation (2)} \qquad F_{drag} = \frac{W^2}{4\,\rho_{air}\,A_{wing}\,v^2}$$

One particularly interesting thing about this result is that *the drag force varies inversely with the velocity squared*. This plotted in the graph below:

This graph tells us that, once the plane takes off, the drag force gets less and less as the plane goes faster and faster. So, to reduce the drag force and fuel consumption, you simply need to fly the plane faster and faster.

But hold on! This defies common sense. You might be thinking that "there's something missing here" – and you would be right!

The drag equation also indicates that we could reduce the drag force as low as we like by simply making the wing area larger and larger. But, aside from the practical engineering challenge of building a plane with enormous wings, there is a fundamental problem here too.

Drag resulting from producing lift

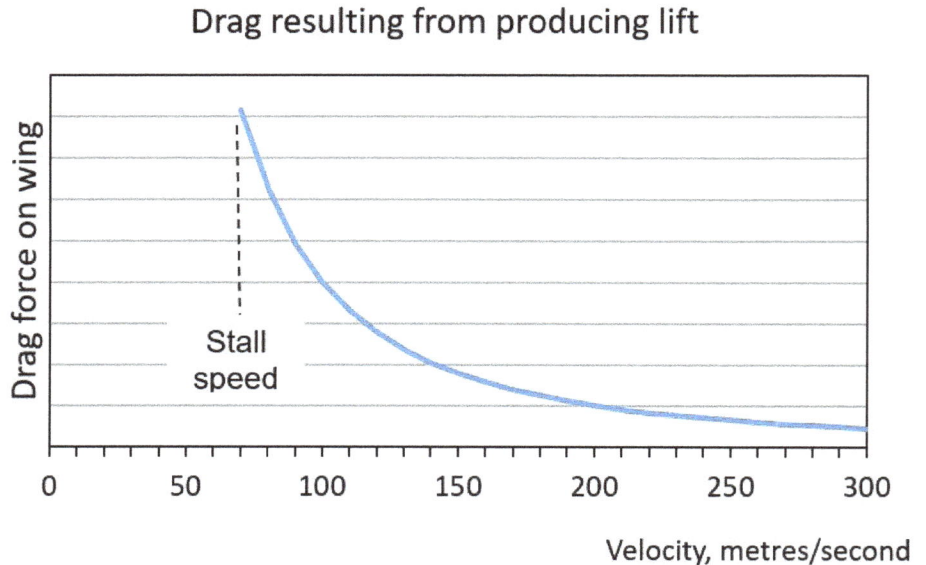

We have only considered the drag force produced in providing lift to support the weight of the airplane. But the plane is flying through the air, so there is also air resistance. There are several types of air resistance, and most sources of air resistance can be greatly reduced by streamlining the shape of the aircraft. But there is one particular source of air resistance that cannot be minimised in this way. This is air resistance due to the viscosity of the air as it flows along the surface of the wing. This creates another source of drag, and this drag force *increases with the velocity squared*.

$$F_{drag} \text{ due to air resistance} = \tfrac{1}{2}\,C_D\,\rho_{air}\,A_{wing}\,v^2$$

Where C_D is the "drag coefficient" due to air resistance

The *total drag force* is the sum of the drag force resulting from the wing generating lift *and* the drag force due to air resistance.

$$\text{Equation (3)} \qquad \text{Total drag force } F_{drag} = \frac{W^2}{4\,\rho_{air}\,A_{wing}\,v^2} + \tfrac{1}{2}\,C_D\,\rho_{air}\,A_{wing}\,v^2$$

10

This creates a fundamental dilemma. To minimise drag resulting from producing lift, we could simply increase the wing area *or* increase the speed of the plane *or* fly at lower altitude (higher air density ρ_{air}). ***But each of these measures increases the air resistance drag!***

So, let's plot a graph showing what happens to the two sources of drag once the plane takes off and increases in speed:

- The drag force resulting from the production of lift (solid blue line) ***gets less at higher velocity***.

- The drag force due to "skin friction" air resistance (dashed blue line) ***increases markedly at higher velocity***.

- The total drag (adding both drag forces) has a minimum value at a particular aircraft velocity. This is the ideal cruise velocity at which the aircraft operates most efficiently. It is interesting that, at this velocity, the two sources of drag are equal.

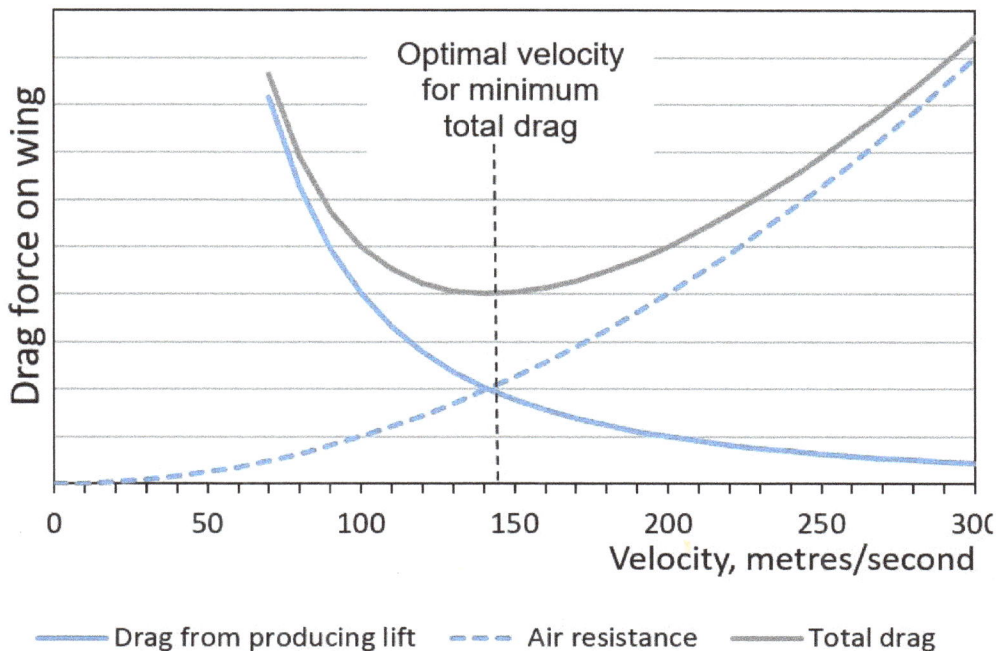

You could say that this optimal cruise velocity is the "sweet spot" for the aircraft to operate.

By applying a mathematical technique called "differentiation"[Note 1] to Equation (4), we can determine the optimal velocity for an aircraft with a particular wing area flying at a particular altitude (air density). What we find is the optimal velocity for minimum drag is:

$$\text{Equation (4)} \qquad \text{Optimal velocity (for minimum drag)} = \frac{W^{1/2}}{2^{1/4}\ C_D^{1/4}\ A_{wing}^{1/2}\ \rho_{air}^{1/2}}$$

This equation tells us a lot about how airplanes should be designed and flown. If you were designing a commercial jetliner, you would probably want it to cruise at a speed of about 250 metres/second (as fast as possible, without getting too close to the speed of sound), and at an altitude of about 10,000 metres (which determines the air density ρ_{air}). Based on the size of the plane, payload and other factors, you would estimate the weight **W** of the aircraft (with fuel and passengers/cargo). You would estimate the value of the drag coefficient C_D. You could then determine from Equation (4) exactly how big the wings need to be (A_{wing}) for the aircraft to cruise at its "sweet spot" at the desired speed and altitude.

Modern jetliners are designed to cruise at an altitude of about 10,000 metres, at which air density is about one-quarter that at sea level. Of course, a plane does not always fly at its cruising altitude. Most airports are near sea level or at relatively low elevation. So, how does the optimal velocity differ between flying near sea level and at 10,000 metres elevation?

As the air density ρ_{air} reaches one-quarter of its value near sea level, the square root of the air density reduces by half, so the equation above tells us that the optimal speed for the plane at cruising altitude should be twice its speed near sea level. This works out well, since airplanes naturally take off (and land) at low altitude and low speed, and cruise at high altitude and high speed. So, pilots gradually increase speed after take-off, as a plane gains altitude, to "track" its most efficient operating point. Furthermore, as we have already seen, the plane loses weight as it burns fuel, so the plane should gradually drift to higher altitude during the cruise phase of the flight.

Let's see how this looks on the graph of drag forces versus speed.

- At low altitude (blue curves), the air density is high, which helps the wings to generate lift force. The drag resulting from generating lift is relatively small (solid blue line), but the air resistance quickly increases at higher speeds (dotted blue line).

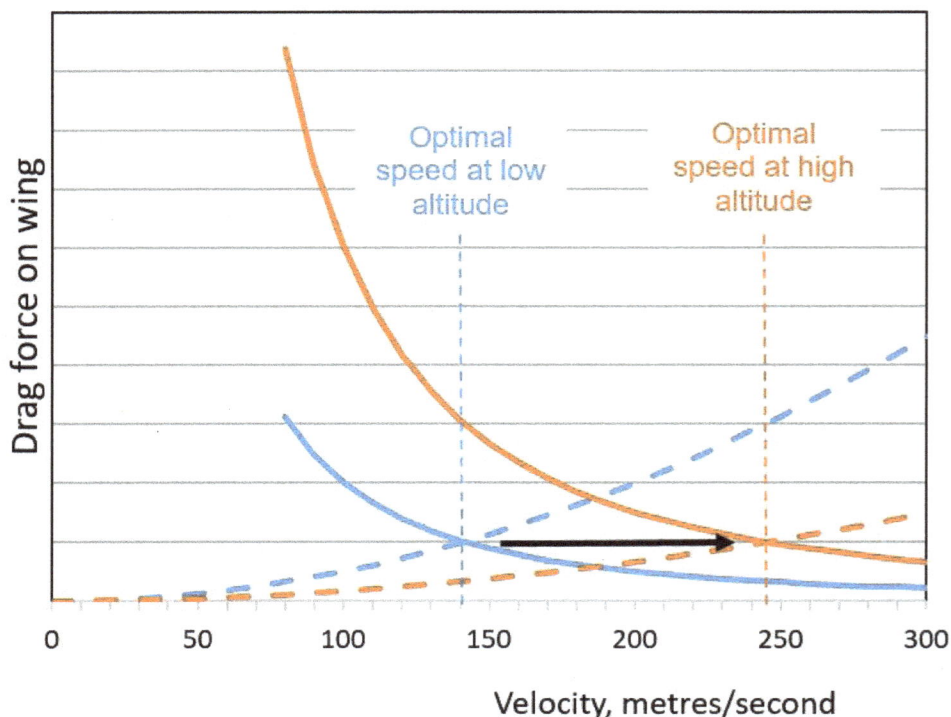

- At high altitude (orange curves). the air is less dense and it is more difficult for the wings to generate lift. To maintain the lift force, the pilot needs to increase the wing's angle of attack, so that the drag force resulting from producing lift increases (solid orange line). On the other hand, as air becomes less dense at high altitude, the air resistance drag (dashed orange line) does not rise as quickly as the speed increases.

The point where the solid and dashed lines cross, at which total drag is a minimum, shifts to higher velocity as the plane gains altitude.

You might be surprised to note that **the total drag force is the same at low altitude and high altitude – providing that the aircraft gains speed at higher altitude to keep within the "sweet spot"**.

In fact, **if the plane flies at its optimal velocity, the total drag force depends only upon weight of the airplane**. The total drag force for a plane flying at its optimal speed can be found by substituting Equation (3) into Equation (4), which gives an astoundingly simple result:

Equation (5)	Total drag at optimal velocity $= [\dfrac{C_D}{2}]^{1/2}$ W

The drag force – and the energy required to move a plane through the air – at the optimal speed depends only on its weight and the "skin friction" drag coefficient. This raises a question that has intrigued me for a long time – is there a theoretical minimum energy requirement for a plane to fly under optimal conditions? And if there is, how would an aircraft be designed to approach the maximum possible efficiency?

I have thought about this a great deal, and provide my analysis and conclusions in the following section.

A fundamental limit to the efficiency of air transport

Let's assume that we build the plane to be as aerodynamically efficient as it could possibly be, setting aside any consideration of cost, inconvenience or impracticality. We do not wish to constrain ourselves by the limitations of technology or materials. So, for example, we could construct the wings to be infinitely long and to taper at the ends. This would reduce the adverse effect of vortices at the wingtips to an insignificant level.

To reduce air resistance, we can eliminate the fuselage, so that all passengers and cargo are carried within a "flying wing". In fact, a number of aircraft have been built along this design strategy. Most were experimental prototypes, This approach was explored in Germany's Gotha 229 jet-powered flying wing fighter bomber in 1945, by Northrup Company in the 1940s with the experimental XP-79 flying wing fighter and XB-35/YB-49 flying wing bomber, and was adopted for the B2 bomber in the 1980s (and reportedly will be used in the US Air Force's next generation B21 bomber). For such "flying wings", the only significant area for air resistance is the top and bottom surface of the wing.

B-2 bomber. US Air Force

In principle, it would be possible to reduce or eliminate all air resistance – except for "skin friction" (or "viscous drag"). As we have seen, air must flow over the wing to generate lift, and – in doing so – the air exerts a viscous force pulling back on the wing. The viscous drag force arises from the difference in velocity between the wing surface and air molecules that hit it. As air travels distance **X** along the wing chord, it produces a "boundary layer" of slow-moving air immediately along the wing surface. Further back along a wing, the boundary layer gets thicker and thicker, partly "insulating" the still air from the moving wing. While the viscous drag force continues to increase as air moves further back along the wing, the **additional drag gets less and less** (the drag per unit area increases with the square root of **X**).

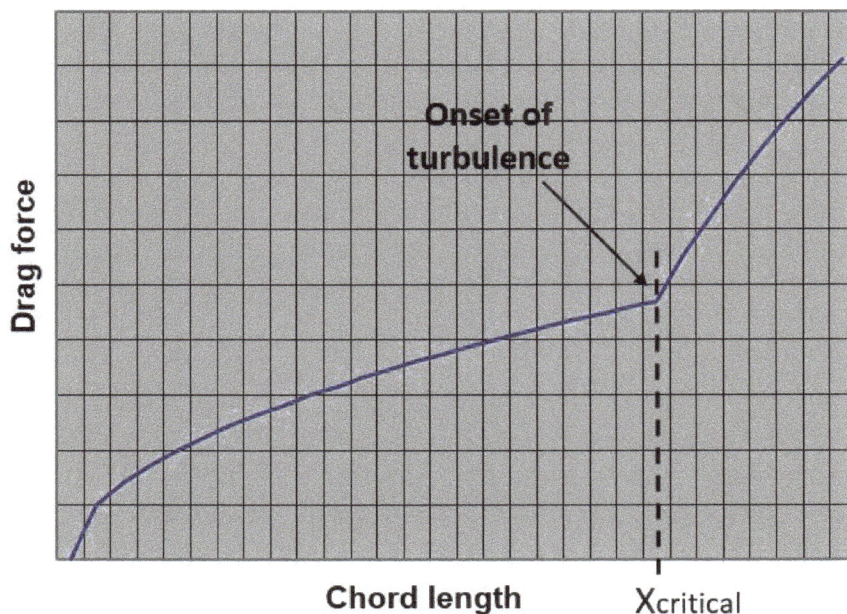

13

That's great, you might think, we could just build the wings really wide, with a huge wing chord. But this wouldn't work, because at some critical distance $X_{critical}$ along the wing, the boundary layer becomes increasingly unstable, until surface roughness, vibration or random fluctuations triggers a transition to "turbulent flow". At this point, drag increases rapidly. The critical distance $X_{critical}$ is determined by the viscosity of the air, the density of the air, the speed of the aircraft and the "critical Reynolds Number" R_c. The Reynolds Number has been measured for air flowing over a flat smooth wing. The results vary with the smoothness and angle of the wing, but lie within the range of 100,000 to 1,000,000.

So, to reduce viscous drag to the lowest possible lever, we should somehow keep the boundary layer at its maximum thickness – the thickness it has *just before* the onset of turbulent flow. In fact, a number of researchers have investigated "active laminar flow control" to do exactly this (by applying suction to the wing surfaces through microscopic holes to prevent the boundary layer from growing to a thickness at which it becomes unstable). None of these techniques seem to have been applied successfully, but let's say that the technology for active laminar flow control could be developed and applied. In this case, the viscous drag would be:

$$\text{Skin friction drag at optimal boundary layer thickness} = \left[\frac{0.66}{R_c^{1/2}} \right] (1/2 \; \rho_{air} \; v^2) A_{wing}$$

This means that, under these idealised conditions, the "drag coefficient" is $0.66/R_c^{1/2}$. Substituting this into Equation (5) gives the total drag force under idealised conditions (flying wing design with active laminar flow control) at optimal velocity:

$$\textbf{Total drag at optimal idealised conditions} = \left[\frac{0.57}{R_c^{1/4}} \right] W$$

Taking the critical Reynolds Number as 1,000,000, at the top of the quoted range, gives a lift/drag ratio of about 55, which is about three times the aerodynamic efficiency of a Boeing 747. This corresponds to a minimum energy requirement of 180 kilojoules per tonne (mass of plane, cargo and crew) per kilometre travelled. I argue that this would be the best possible "aerodynamic efficiency" of an ideal airplane. Of course, actual fuel consumption also depends on the efficiency of the engines. In an idealised world, with no limitations of technology or materials, we can imagine that aircraft engines might approach 100% efficiency. In this case, the absolute minimum fuel consumption would be about 0.5 litres of jet fuel per 100 kilometres flown per tonne. This is roughly about one-eighth of the fuel consumption of modern jetliners on long-haul flights.

So, in principle, there is still considerable scope for future technological improvements to reduce fuel consumption for air transport. In recent decades, significant progress has already been achieved in increasing the efficiency of jet engines (although they are still nowhere near 100% efficient).

Fuel consumption depends on the total weight of the aircraft, which primarily consists of the aircraft itself, with passengers and their luggage comprising only perhaps 10% of the take-off weight. Passenger aircraft of recent design and construction (like the A380 and Boeing 787) make extensive use of carbon fibre composites to reduce the weight of the airframe, and pack in as many passengers as possible (which, as most of us know, has its drawbacks for passengers enduring cramped conditions on long flights).

Another way to reduce the weight of an aircraft is to not have *any* passengers or crew, and this is one of the main drivers for the development of drones. For military reconnaissance and aerial surveys, why carry a human pilot and observer (weighing at least 100 kilograms with seat and other gear), when the same function can be performed by a computer, cameras and sensors weighing less than a kilogram? Such applications can be performed by small, lightweight drones (operated remotely or even autonomously), with potentially huge reductions in cost and fuel consumption. However, replacement of a human pilot and crew would yield only marginal weight reductions in commercial airliners.

Bear in mind that improvements in energy efficiency often become self-limiting. As aircraft have become more fuel-efficient and cheaper to operate, the number of people flying long distances for holidays and work has grown enormously. While the energy required to transport each tonne-kilometre of passengers and cargo has been reduced through improved aircraft design, the volume of passengers and cargo has grown even faster, and so have the distances travelled. As extraordinary as it might seem, my son-in-law commutes between Australia and Africa on an eight-week fly in-fly out cycle. Over the past year, Tim would have flown something like 200,000 kilometres. Over the past two years, he would have flown the distance from the Earth to the moon. This would have been unthinkable a few decades ago, but is increasingly becoming the "new normal" for many working professionals, although this lifestyle leads to very high fuel consumption and can be very socially disruptive of family life and social relations.

Notes
(1) To find the velocity **v** which gives minimum drag force, we take the derivative of the drag force with respect to velocity **v**, and set this equal to zero (since the curve of drag force versus velocity is flat, or has zero slope, at its minimum value). Considering all other factors (the weight **W**, air density ρ_{air}, A_{wing} and C_D) as constant, we can rewrite Equation (3) in terms of velocity **v** and two new constants **A** and **B**.

$$\text{Total drag force} = \frac{A}{v^2} + Bv^2 \qquad \text{Where } \mathbf{A} = \frac{W^2}{4\,\rho_{air}\,A_{wing}} \text{ and } \mathbf{B} = \tfrac{1}{2}\,C_D\,\rho_{air}\,A_{wing}$$

Taking the derivative, and setting it equal to zero gives: $-2\,\dfrac{A}{v^3} + 2Bv = 0$

Re-arranging to solve for the optimal velocity gives: $v = \left[\dfrac{A}{B}\right]^{1/4}$

Substituting this optimal velocity back into the equation for drag force gives:

$$\text{Total drag force at optimal velocity} = 2\,A^{1/2}\,B^{1/2}$$

When we substitute the expressions for **A** and **B**, the air density ρ_{air} and wing area A_{wing} cancel out, leaving a remarkably simple Equation (5) for the drag force at the optimal velocity.

4. Helicopters

I have previously referred to the "recoil" force produced by a stream of water shooting out of a fire hose, and suggested that it might even be possible (but not very practical) to suspend an aircraft in the air by shooting a stream of water downwards. As I have discovered, this is exactly what is done at some amusement parks, where people are fitted with a backpack connected to a hose carrying a rapidly-flowing stream of water. The backpack directs the flow downwards, lifting the person into the air. A good 8-minute video ("Jetpack Rocket Science") showing how this works, and explaining the principle can be found at: https://www.youtube.com/watch?v=Hx9TwM4Pmhc

Airplanes and helicopters fly at high altitude, and do not have ready access to large volumes of water that can be propelled downwards. But they do have access to large volumes of air, so airplanes and helicopters work by accelerating the air downwards. Aircraft use their wings to deflect the air downwards as they move horizontally. Helicopters use a large rotating propeller (rotor) to impart downwards momentum to the air. The process of accelerating the air downwards generates an upwards lift force which supports the weight of the aircraft.

However, by accelerating air downwards, we are imparting kinetic energy to the air, as well as momentum. To minimise the energy needed to stay aloft, we want to impart the minimum amount of kinetic energy to the air, while imparting sufficient momentum to support the weight of the aircraft. To do this, *we want to accelerate the maximum amount of air, to the minimum velocity necessary*. In the case of a helicopter, the power requirement is kept to a minimum by using a large diameter rotor to accelerate air over a very large area.

The circular area swept by the spinning blades of a helicopter is called the "swept area" of the rotor, or the "rotor disc".

Although we talk about the spinning rotor as accelerating the air downwards, this cannot occur within the thin disc-shaped area swept by the rotor. We normally think of the air as being "compressible", but at the low pressures and velocities that we are considering here, air acts very much like an *incompressible* fluid.

Since the volume of the air stays nearly the same as it passes through the rotor disc, air cannot leave the bottom of the rotor disc faster than it enters at the top. What happens is that *the spinning rotor reduces the pressure just above the rotor disc*, sucking air downwards towards the disc. Well above the spinning rotor blades, the air is stationary, and its downwards velocity is zero. Reduced pressure above the rotor disc pulls the air downwards, until it enters the rotor disc at some velocity v_{rotor}.

Simultaneously, *the spinning rotor increases the pressure just below the rotor disc*, further accelerating the air. Air passes through the rotor disc at velocity v_{rotor}, and then, as it moves below the rotor (and to lower pressure), it is accelerated to a final velocity v_2.

As it turns out, and as I derive in a following section, the conservation of momentum requires that the velocity of air moving through the disc v_{rotor} is the average of its initial and final velocities. Since the initial velocity of the air is zero, and its final velocity is v_2, the velocity of the air moving through the disc is the average of the initial and final velocity, which is $\frac{1}{2} v_2$.

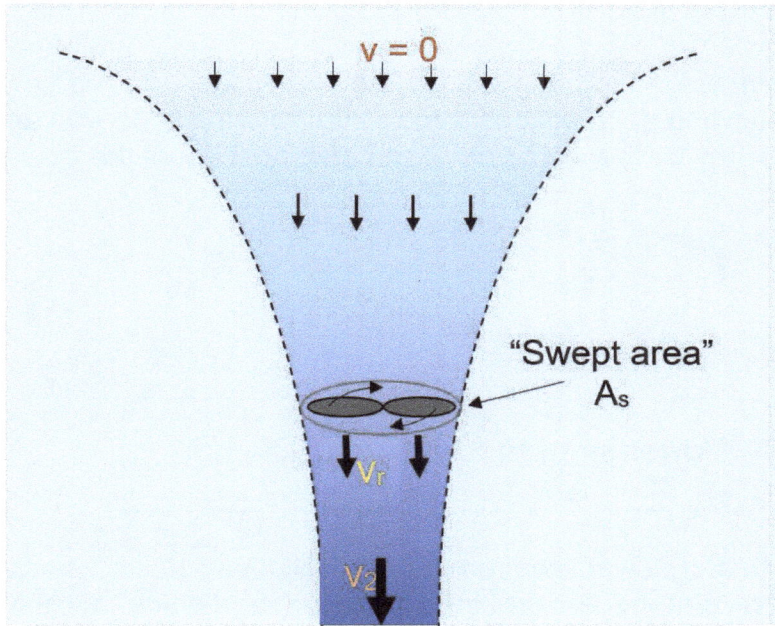

v = 0

"Swept area" A_s

v_r

v_2

Consider the volume of air flowing each second through the rotor disc of a helicopter. The volume of air is given by the cross-sectional "swept" area A_s of the rotor disc, multiplied by the velocity of air v_{rotor} moving through the rotor disc.

Volume of air/second $= A_s\, v_r$

The mass of air flowing through the disc each second is given by the volume multiplied by the density of air.

Mass flow $= \rho_{air}\, A_s\, v_r$

Since the velocity of air flowing through the rotor disc v_{rotor} is half the final velocity v_2 imparted to the air, the mass of air flowing each second through the rotor disc is $\frac{1}{2}\,\rho_{air}\, A_s\, v_2$.

The upwards lift force produced by the spinning rotor blades is equal to the rate at which momentum is imparted to the airflow, which is the mass of air flowing through the rotor disc multiplied by its increase in velocity.

Upwards force on rotor $= (\, \tfrac{1}{2}\,\rho_{air}\, A_s\, v_2\,)\,(\,v_2 - 0\,)$

mass of air/second — change in velocity of airstream

When the helicopter is hovering or flying horizontally, the upwards force is equal to the weight of the helicopter **W**. Substituting in the equation above, we get:

Weight of helicopter, $W = (1/2\ \rho_{air}\, A_s\, v_2)\,(v_2)$

Combining terms, this gives:

Equation (1)   Weight of helicopter, $W = 1/2\ \rho_{air}\, A_s\, v_2{}^2$

This is great. We simply need to rotate the rotor blades fast enough to keep the helicopter in the air.

Note however that, in accelerating the air to velocity v_2, the rotor imparts kinetic energy to the airstream. To keep the helicopter suspended in the air, we must continually provide power to the rotor to replace the energy transferred to the airstream as kinetic energy.

We have seen before that the kinetic energy of any moving body is given by its mass and velocity squared, as follows:

Kinetic energy of mass **m** moving at velocity $v = \frac{1}{2} mv^2$

In this case, the "moving body" is the mass of air that is accelerated each second to its final velocity $\mathbf{v_2}$. We know the mass of air flowing each second through the rotor disc, so we can readily calculate the gain of kinetic energy by the airstream – and this must equal the power provided to the rotor.

$$\text{Power input to rotor} = \frac{1}{2} \underbrace{(1/2 \, \rho_{air} \, A_s \, v_2)}_{\text{mass flow}} v_2^2$$

This gives:

Equation (2)	Power input to rotor = $\frac{1}{4} \rho_{air} \, A_s \, v_2^3$

Equations (1) and (2) relate the weight and power requirement of a helicopter to the velocity imparted to the airstream v_2. We can re-arrange and combine these two equations to eliminate v_2. This allows us to determine the power required to support the weight **W** of a helicopter. The result is:

$$\text{Power required} = \frac{1}{\sqrt{2}} \frac{W^{3/2}}{\rho_{air}^{1/2} A_s^{1/2}}$$

This can also be written as:

Equation (3)	Power required = $\frac{1}{\sqrt{2}} \left[\frac{1}{\rho_{air}^{1/2}} \right] \left[\frac{W}{A_s} \right]^{1/2} W$

Thus, the power required for a helicopter to hover varies directly with the weight **W** of the helicopter, but also with the ratio **W/A**, the weight per unit area swept by the rotor. This is the **minimum power** required to accelerate the air downwards, and since a rotor will always have some aerodynamic losses, the actual power requirement will always be greater than this.

For minimum fuel consumption, the helicopter should have the largest possible rotor diameter.

As an example, consider the common R22 small helicopter. The specifications are given as follows (at http://en.wikipedia.org/wiki/Robinson_R22):

Rotor diameter	7.67 metres
Rotor disc area	46.2 m²
Empty weight (mass)	389 kg
Maximum take-off weight (mass)	635 kg
Powerplant	93 kilowatt piston engine

R22 helicopter, Wikimedia Commons,
Photo by James from Cheltenham

Using the equation above, and substituting the values for rotor disc area **A**, weight **W** and air density ρ_{air} of 1.2 kg/m³ at sea level, we can calculate the minimum power required for take-off with the maximum load is 46,550 watts (46.5 kilowatts). The engine must be sufficiently powerful to meet this minimum power requirement, as well as overcome skin friction drag on the rotor blades and other aerodynamic losses. The rated 93 kilowatt power rating of the engine is about twice the minimum power requirement.

The equation that we derived shows that the power requirement increases:

- as the density of the air reduces (at higher altitude),

- as the "rotor loading" (weight per unit area swept by the rotor) increases,

- as the weight of the helicopter and its cargo increases.

Consequently, to reduce the power requirement for a helicopter, it is important to have the largest feasible swept area by the rotor, and the minimum weight of the helicopter.

One technological revolution that is underway is the application of electronic sensors and control systems to unmanned aircraft. So, for example, if you are using a helicopter for surveillance (for forest fires, police work, or for soldiers to locate enemy positions), it is no longer essential to have humans on-board. In many such cases, you can replace a 75 kilogram human observer with a 75 gram video camera and transmitter – with one-thousand times less weight! At the same time, lithium battery technology offers much greater energy storage densities than conventional batteries, so very small, remotely-operated helicopters powered by batteries and electric motors are now available, and can even be bought as toys! Furthermore, electronic control software enables sophisticated and complex manoeuvres to be undertaken by relatively unskilled operators.

A four-minute video showing the amazing manoeuvres that can be performed with a remote-controlled helicopter model can be viewed at:
https://www.youtube.com/watch?v=WdEWzqsfeHM

The propeller "mystery"

When you look at a helicopter rotor, or a propeller or a wind turbine (a propeller operated in reverse), it immediately becomes apparent that the blades are relatively thin. The rotor blades occupy only a small fraction of the total area of the rotor disc. I have long been intrigued by why this should be so. Wouldn't the rotor work better, I thought, if the rotor blades were wider (longer chord) or if there were more blades?

I have also been intrigued by another question. The blades of a helicopter rotor (or a propeller, or a wind turbine) are simply wings that rotate in a circular path. The rotor blades of a helicopter should act exactly like the wings of an airplane, so how can we reconcile the equation that we derived for the lift of an airplane wing with the equation for the lift of a helicopter rotor?

The answer lies in the fundamental difference between an airplane wing and the blades of a rotor. An airplane wing travels through still, "virgin" air. As the wing passes through stationary air at velocity **v**, the wing deflects the air downwards. As we have seen, air immediately next to the wing surface follows the contours of the wing and acquires a downwards velocity **vθ** (where **θ** is the angle of tilt of the wing, or "angle of attack"). Further out from the wing surface, the downwards velocity imparted to the airstream become progressively less. As described earlier, the disturbance of the airstream extends an "effective distance" which is roughly equal to the chord of the wing.

Unlike an airplane wing, the blade of a helicopter rotor follows in the wake of another blade which previously passed through the same air. If the blade follows its colleague too closely, it hits air that has already been disturbed and already has significant downwards velocity. In this case, the blade will not be able to operate at full effectiveness. The job of the rotor blade is to deflect the air downwards - but if the air is already moving downwards, the blade becomes less effective or redundant. Ideally, rotor blades should be spaced so that, during the time it takes the second blade to reach the same place, the "disturbed air" will have passed through the rotor disc. Then, the blade will encounter air which has experienced little disturbance from the previous blade passage.

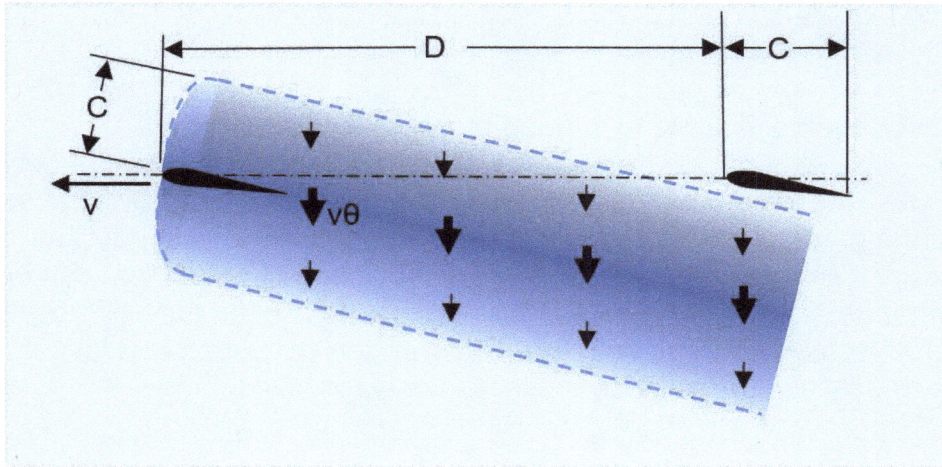

So, let's see what condition must apply for each rotor to pass through relatively "undisturbed" air. Let's assume that the distance between rotor blades is **D** along a circular path at some radius from the rotor hub), and that the rotor blades are rotating at velocity **v**.

First, let's consider what happens as the first rotor blade passes a fixed point. The air above and below the blade, extending up and down a distance equal to the chord length **C**, is given a downwards velocity **vθ**. The "wake" behind the blade will have an effective height **2C** and, at its mid-line, will be moving downwards at velocity **vθ**.

Imagine that we start a timer as the first rotor blade passes this point. The time for the following rotor blade to reach the same point is **D/v**. During this time, the airstream will move downwards by distance **(D/v)(v θ)**. If this distance is greater than the chord length **C**, then the "disturbed" air will have passed through the plane of the rotor before the next blade arrives.

Thus, each rotor blade hits undeflected "virgin" air if **(D/v)(v θ)** is greater than **C**. So, the ratio of the blade chord to the distance between blades, **C/D**, is equal or greater than the "angle of attack" of the rotor blades. This means that the ratio of the area of the blades to the area of the rotor disc is equal to the "angle of attack" of the blades.

$$\frac{\text{Area of rotor blades, } A_w}{\text{Swept Area of rotor, } A_s} = \theta \qquad \text{Condition for rotor blades to hit "un-deflected" air}$$

For an aircraft wing, a typical angle of attack is about five degrees (roughly, one-tenth of a radian), so we would expect that the rotor area would be roughly one-tenth the circular area swept by the rotor.

If rotor blades are wider than they need to be, the lift would increase only marginally, but the larger rotor surface area would greatly increase skin friction drag and power losses.

The same reason explains why the biplane design used in nearly all aircraft up to the 1920s (with two wings, one mounted above the other) was abandoned and replaced by a single wing. Unless the two wings are far apart, each "feels" the airflow that has already been deflected by the other wing.

Personal jetpacks, backpack helicopters and jet-powered hoverboards

Back in the 1960s, many people believed that we would soon be strapping rocket-powered packs on our backs and soaring through the air. To be literally able to soar and hover like a bird would be a truly amazing experience, and indeed, such rocket packs were developed and trialled by the military, but they didn't catch on.

Like a helicopter, rocket packs develop lift force to support the weight of the pilot by accelerating fluid downward. For rocket packs, the rocket fuel itself is vaporised and shot out of a nozzle at high velocity. But here's the problem. To develop lift force with the minimum use of energy, we need to direct the greatest amount of mass downwards at the lowest possible velocity. In a rocket pack, you are limited to the relatively small mass of rocket fuel that can be carried on the rocket pack strapped to the pilot's back. Most rocket packs used pure hydrogen peroxide, which is a potent rocket fuel (a far cry from the 3% hydrogen peroxide used for bleaching hair), but even so, the flying time of these rocket packs was limited to one or two minutes.

To get around this limitation, development shifted to "jet packs", utilising small jet engines to produce lift by directing their exhaust downwards. The mass being accelerated downwards (jet exhaust) consists mostly of air, which is drawn from the surrounding atmosphere (as in a conventional helicopter). But this is still not very energy-efficient, since the inlet area of a jet engine that a person can strap to their back is quite small.

Let's say that the inlet of a small, portable jet engine is 120 millimetres in diameter. This means that all of the air being accelerated downwards must pass through an opening with an area of 0.017 square metres. Let's say that the total mass of the pilot and the jetpack is 120 kilograms (giving a weight of 1,200 Newton). At low altitude, air has a density of 1.2 kilograms per cubic metre. According to the "helicopter equation" (Equation 3), to support the weight of the pilot and jetpack, a minimum 200,000 Joules of work must be expended every second to accelerate the jet exhaust downwards. If the jet engine converts one-quarter of the fuel energy into kinetic energy of the exhaust gases, the engine would need to consume 1.5 litres of jet fuel every minute. (about ten times the rate of fuel consumption to drive your car along the highway).

The company Jetpack Aviation (http://jetpackaviation.com) is currently seeking investors to begin commercial production of its model JB-9, which is stated to be the first true personal jetpack. A two-minute video showing a pilot flying a JB-9 prototype over a pond (you can see the effect of the downwards jet exhaust on the water surface) can be viewed at: http://jetpackaviation.com/the-jumpjet/jb-9/ .Fuel consumption of the JB-9 is reported to be 4 litres per minute, giving ten minutes of flight with a 38-litre tank of jet fuel.

Other inventors and entrepreneurs have also developed (and some are even selling) jet-powered sit-on or strap-on helicopters, and jet-powered stand-on hoverboards. Some use the downwards exhaust of a jet engine to produce lift, while others use a jet engine to power a series of small rotors (which increases the area through which air is accelerated, and should be more efficient). A ten-minute video showing the diversity of such devices may be viewed at: https://www.youtube.com/watch?v=588du55BcAY

As this video shows, such backpack or stand-on helicopters can be very small and compact, yet support the weight of a person. Lift is achieved with a very small rotor area by using powerful engines. The "price to pay" for such compact rotors is high fuel consumption and, since only limited fuel can be carried, flights of limited duration.

Human-powered helicopters

The exact opposite approach is to develop a helicopter powered entirely by its human operator. Such a helicopter would need to be designed and built so that it could hover with an absolute minimum power requirement. In fact, it had been debated for some time whether it would even be possible to build a helicopter powered solely by the physical exertion of its human operator. A human-powered winged aircraft (the Gossamer Condor) had been demonstrated in 1977, and two years later, an improved model (the Gossamer Albatross) was flown (pedalled) across the English Channel (for a video, see: https://www.wired.com/2010/08/0823gossamer-condor-human-powered-flight). However, a human-powered helicopter proved to be an even more difficult challenge.

To settle whether such an achievement would be possible, a Human Powered Helicopter Competition was sponsored by the American Helicopter Society (AHS). A modest prize was initially offered in 1980 for any team that could fly a human-powered helicopter for at least one minute at a height of 3.3 metres. For over 30 years, no one could meet the challenge, and many believed that the goal was unattainable. The prize was increased to $250,000 in 2009, and in 2013, the competition was won by a team of students from the University of Toronto.

The winning entry was called the "AeroVelo Atlas". it had four enormous rotor blades, giving a total "swept area" of 1,280 square metres, to minimise the power requirement (as would be expected from the "helicopter equation"). The helicopter was constructed with a carbon fibre frame and other lightweight composite materials to keep the weight of the craft to an absolute minimum (55 kilograms). The "AeroVelo Atlas" was flown by Todd Reichert, a racing cyclist and pilot who trained to provide an initial 1.1 kilowatt power surge for take-off, and then maintain 600 watts power to hover for one minute.

The "helicopter equation" indicates that such a craft would require the pilot to produce a power output of at least 800 watts. However, at the low altitude flown (much less than the diameter of the rotors), the power requirement was reduced by the "ground effect". This arises for craft flying just above the ground, so that air accelerated downwards by the rotor (or wing) cannot readily escape from beneath the rotor, increasing the air pressure there.

A short video showing the flight of the AeroVelo Atlas can be viewed at:
https://www.youtube.com/watch?v=syJq10EQkog

Derivation of the fluid velocity passing through a rotor disc
(optional for those who are interested in the mathematical derivation)

Imagine the airflow passing through the rotor disc of a helicopter rotor, a propeller, a fan or a wind turbine. The rotor may be imparting kinetic energy to the air (as in a helicopter rotor, propeller or fan) or extracting kinetic energy from the air (as in a wind turbine).

In the case of a fan or propeller, the airflow speeds up, and decreases in area as it approaches and then leaves the rotor disc. Note that the only object exerting a force on the airstream is the turbine located within the rotor disc. The airflow will look like this (with increasing velocity shown as a darker shade of blue):

Each second, a more-or-less cylindrical volume of air (shown as a thin disc) approaches the rotor disc at an initial velocity v_o. When it reaches the rotor disc, it has been accelerated to velocity v_r. It achieves a final velocity v_2 at some distance downstream from the rotor.

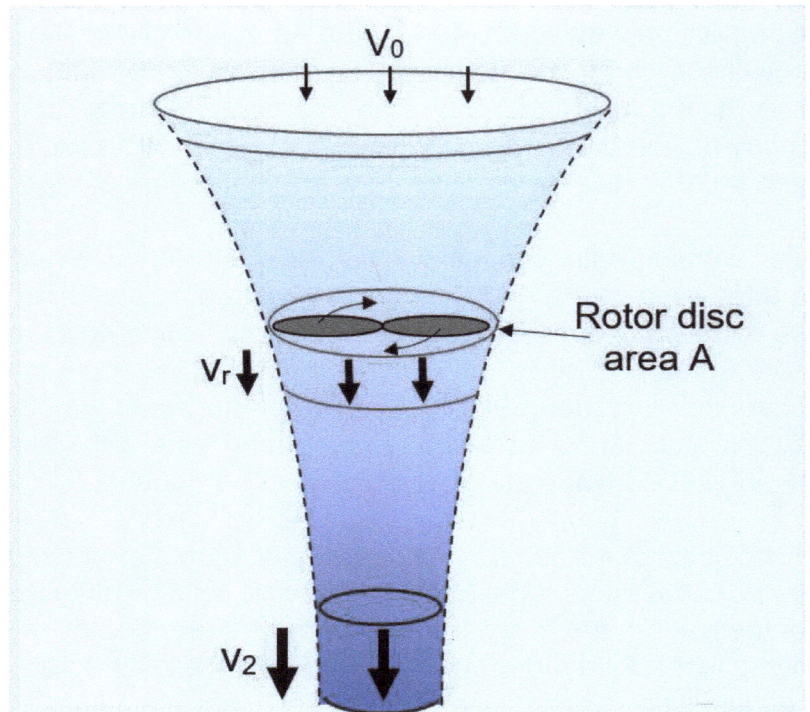

The volume passing through the rotor disc is given by the area **A** of the rotor disc multiplied by velocity v_r. The mass flowing through the rotor disc each second is its volume multiplied by the density of the fluid **ρ**. So, the mass flowing through the rotor is **ρAv_r**.

The kinetic energy gained by the fluid passing through the rotor disc each second is the difference between the final kinetic energy (at velocity v_2) and its initial kinetic energy (at velocity v_o). This kinetic energy must be provided by the turbine within the rotor disc, and is the mechanical power that must be supplied to the rotor.

$$\text{Power input} = \tfrac{1}{2}(\rho A v_r)\, v_2^2 \; - \; \tfrac{1}{2}(\rho A v_r)\, v_o^2$$

$$\underbrace{\qquad\qquad}_{\substack{\text{Final kinetic}\\ \text{energy}}} \qquad \underbrace{\qquad\qquad}_{\substack{\text{Initial kinetic}\\ \text{energy}}}$$

Combining terms to simplify the equation, we get:

Equation (1) $\qquad\qquad \text{Power input} = \tfrac{1}{2}(\rho A v_r)[\, v_2^2 - v_o^2\,]$

As well as imparting kinetic energy to the fluid, the turbine also increases its momentum (the mass of fluid multiplied by its change in velocity). The change in momentum is provided by the force applied to the fluid volume as it passes through the rotor disc.

Each second, fluid passing through the rotor disc travels distance v_r. The rotor disc does work in applying this force over distance v_r each second. Thus, the power transferred to the fluid by the rotor is given by:

$$\text{Power input} = \underbrace{(\rho\,A\,v_r)\,[\,v_2\,-\,v_o\,]}_{\substack{\text{Force exerted} \\ \text{on fluid}}}\,v_r$$

Simplifying, we get:

Equation (2) $\qquad \text{Power input} = (\rho\,A\,v_r^2)\,[\,v_2\,-\,v_o\,]$

Equations (1) and (2) both relate to the power input required to accelerate the airstream from velocity v_o to velocity v_2. Equation (1) was derived based on the conservation of energy, while Equation (2) was derived based largely on the conservation of momentum. Setting these two equations equal to each other, we get:

$$\underbrace{\tfrac{1}{2}\,(\rho A v_r)[\,v_2^2 - v_o^2\,]}_{\substack{\text{Power, given by} \\ \text{Equation (1)}}} = \underbrace{(\rho\,A\,v_r^2)\,[\,v_2 - v_o\,]}_{\substack{\text{Power, given by} \\ \text{Equation (2)}}}$$

Both sides of the equation contain the term $\rho A v_r$, which can be factored out. This leaves:

$$\tfrac{1}{2}\,[\,v_2^2 - v_o^2\,] = v_r\,[\,v_2 - v_o\,]$$

The term $[\,v_2^2 - v_o^2\,]$ can be factored into $(v_2 - v_o)(v_2 + v_o)$. This gives:

$$\tfrac{1}{2}\,(v_2 - v_o\,)(v_2 + v_o) = v_r\,[\,v_2 - v_o\,]$$

When we factor out the term $(v_2 - v_o)$, which appears on both sides of the equation, we get simply:

$$v_r = \frac{v_o + v_2}{2}$$

This gives us the completely general result that the velocity of the fluid through a rotor disc is the average of the initial and final velocities of the fluid.

5. Nuclear energy: Mankind's saviour or curse for the 21st century?

Introduction

In the post World War II period of my childhood, nuclear energy was seen as the key to the future. It was presented as a source of limitless cheap energy that would power industry, our homes and maybe even our cars. There was a lot of hype, naivety and gung-ho optimism that were characteristic of the time.

The huge promotion and funding of civilian nuclear power was, I think, a cover for a huge military program to further develop and deploy nuclear weapons. The United States (as well as Britain and France, which developed their own nuclear weapons) was becoming locked into a "cold war" with the Soviet Union, which successfully developed and tested nuclear bombs only a few years after the United States.

The race was on to build thousands of nuclear weapons, as well as aircraft, submarines and missiles that could carry them to distant targets. Many of these nuclear bombs had tens or hundreds of times more explosive power than the atomic bombs dropped on Hiroshima and Nagasaki at the end of the WWII. The nuclear arsenals of the United States and Soviet Union had (and still have) the capacity to kill much of the world's population, and to end modern civilisation.

Various scenarios were considered for the consequences of a nuclear war: would survivors be able to survive by adapting simple pre-industrial or hunter-gatherer lifestyles, or would it mean the extinction of the human species? There was no doubt that the consequences of a nuclear war would be an apocalyptical catastrophe for mankind on a scale never seen before.

The survival of mankind depended on the strategy of "Mutually Assured Destruction" ("MAD"), where neither side dared to attack the other, knowing that their country and population would be destroyed by nuclear retaliation.

At the time, I was living in New York City, which was generally believed to be at the top of the list of targets in the event of a nuclear war. I recall reading with great interest a book explaining how to build a fallout shelter (to protect the occupants from radiation produced by a nuclear bomb, assuming you were far enough away not be incinerated or blown apart by the blast). The book had great colour drawings showing how such a shelter could be built in the basement of your house. I eventually realised that this had limited relevance to my circumstances, living on the 13th floor of a twenty-story apartment building.

On the other hand, some of the positive potential for nuclear energy has been realised. Several hundred nuclear power reactors were constructed in a few dozen countries, and provide a significant portion of the world's electrical energy requirement. Unlike other fossil fuel energy sources, production of nuclear energy does not release much carbon dioxide or other greenhouse gases to the atmosphere (although it does produce relatively small amounts of highly radioactive "high-level" nuclear waste, which will pose a hazard for hundreds or thousands of years).

As well as being radioactive, some radioisotopes produced in nuclear reactions are literally "hot". Their heat is used in radioisotope generators to produce electrical power for spacecraft. Although the amount of electricity produced is quite modest, it is generated continuously over many years, providing sufficient power for communication and guidance systems. Without such radioisotope generators, space missions to the outer solar system would not have been feasible.

Knowledge of radioisotopes explains the heat source for volcanoes and movement of continental plates across the Earth's surface, and allows the age of fossils and ancient rocks to be determined. Radioisotopes are also used in medical applications, particularly for diagnosis and treatment of cancers.

Experimental nuclear fusion reactors have been developed to utilise the type of nuclear reactions that power the sun. If development of this technology is ultimately successful, nuclear fusion reactors could provide much of the world's electricity with minimal greenhouse gas emissions, with less production of long-lived nuclear waste, and with less likelihood of being used as a cover for nuclear weapon production. Currently, some billions of dollars of development funding are being devoted to build and test a prototype nuclear fusion reactor, which is intended to demonstrate the potential for nuclear fusion to be a practical, cost-effective source of electrical power. Would these resources better be invested in further development of solar, wind and other renewable energy resources? The answer depends on your point of view.

From one perspective, nuclear energy is one of the greatest scientific developments of the 20th century, and has potential to be the major energy source for mankind in the 21st century. From another perspective, nuclear weapons continue to be the greatest threat to the survival of mankind.

As much as in any other area of science, nuclear energy illustrates that technology itself is moral-neutral. Scientific knowledge and technology provide the capability for society to do things, for good or evil, that couldn't be done otherwise. Some would argue that humanity would be better off if nuclear energy had never been discovered. This view implies that we should work towards phasing out nuclear energy and nuclear weapons, so that the capability to build and operate such systems would fade and eventually be lost.

Others would argue that knowledge of how to use nuclear energy, for useful purposes or for weapons, cannot be "undiscovered". "The "nuclear genie is out of the bottle, and cannot be put back". Instead, they would argue, we should try to maximise its huge potential benefits.

Some fifteen years ago, I watched the movie trilogy "Lord of the Rings", based on the epic fantasy novel by JRR Tolkien. The story is about a ring that confers supernatural powers to the one who wears it. Whoever wears the ring controls the fate of civilisation. The hero Frodo faces arduous trials and dangers from evil forces seeking to gain control of the ring and the power it offers. In the end, Frodo finally holds the ring in his own hands. He is horrified by its alluring and corrupting influence, and flings it into a boiling cauldron of molten rock to destroy it forever. Could the ring be a metaphor for nuclear technology and its potential for good or evil?

6. Discovery of the atomic nucleus

For over 150 years, scientists have had a pretty good understanding of chemical reactions: which substances react with each other, and which don't.

They knew that substances are composed of atoms, the smallest unit of any element. To understand what they means, let's consider a simple "thought experiment". If we take a lump of, say, graphite (carbon) and cut it in half, we would get two smaller pieces of carbon. We could keep cutting it into smaller and smaller pieces, and we would still have microscopic bits of carbon. As pieces got smaller and smaller, eventually we would get a single atom of carbon, and that would be the smallest piece that would still be carbon.

The existence of atoms has been well-established and is the basis of modern chemistry. Even so, when I first studied chemistry in high school, no one had ever seen an atom and it seemed inconceivable that this would ever be possible. However, with the advent of technologies like Scanning Tunnelling Microscopy and Atomic Force Microscopy, we can now actually "see" atoms. If we look at the surface of carbon, for example, we see what looks like fuzzy balls in a regular array of a crystal structure. The "fuzzy balls" are spherical atomic orbitals of outer-shell electrons. The atomic orbital represents the region around the atom in which there is a high probability of finding an electron.

A carbon atom is roughly about 2×10^{-10} metres in diameter. To put this on a scale that we can relate to, imagine that each carbon atom is the size of a fine grain of sand (about 0.1 mm diameter). On this scale, one gram of carbon would cover the entire area of Brisbane City Council (1,367 square kilometres) to a height of about 18 metres.

It is electrons, residing within the "atomic orbitals" of atoms, that are responsible for chemical reactions. Sharing or transfer of electrons between atoms gives rise to chemical bonds, and the chemical properties of each element. Consequently, chemists live in a world of electrons and atomic orbitals, and electrons were the first of the key constituents of atoms to be understood.

Electrons are relatively loosely held in atomic orbitals, and can be dislodged by high temperatures (or by ultraviolet radiation, or impact by other electrons). Soon after the invention of the light globe in the late 19th century, it was discovered that electrons "boil off" the metal surface of a white-hot filament of a light globe and escape to the surrounding vacuum, from where they can be collected on another metal electrode. This led to the manufacture of glass vacuum tubes containing emitter and collector metal electrodes, which were used as electronic amplifiers in early radios (and later, televisions).

By dislodging electrons from atoms, scientists could readily study their properties. In particular, the electrical charge and mass of the electron have been measured to very high accuracy. Electrons have a negative electrical charge, but relatively little mass.

So, while the properties of electrons were already being widely exploited in the fledgling electronics industry, it was clear that electrons were only one of the key constituents of an atom. It was clear that there were others. For one thing, the electrons in an atom account for less than one-thousandth of its mass. So, other particles in an atom must be much more massive than electrons.

Since atoms are electrically neutral, they must contain positively-charged particles to balance the negative charge of electrons. This positively-charged particle was named the "proton".

Each proton has a positive charge that is exactly equal, but opposite to, the negative charge of an electron. For atoms to be electrically neutral, they must contain the same number protons and electrons. Each element has a particular number of protons (the same as the number of its electrons), and this is the "atomic number" of that element.

The simplest element is hydrogen, which has one proton (and, of course, one electron). Hydrogen has an atomic number of one. The next element is helium, with two protons.

Around 1900, Henri Becquerel and Marie and Pierre Currie discovered that certain elements (uranium, thorium, polonium and radium) were radioactive. One type of radioactivity is "alpha radiation", consisting of positively-charged helium atoms moving at extremely high velocity. Such radiation could be detected in Geiger counter detectors because each alpha particle has so much energy that it could blast electrons out of thousands of gas molecules, and this could be detected by causing a pulse of electrical current to flow through the gas.

Early in the 20th century, physicists were trying to establish the structure of atoms. Since they knew that electrons could relatively easily be dislodged from atoms, it was probably reasonable to imagine that the electrons were embedded within a positively-charged matrix, resembling sultanas in a plumb pudding. This was called the "plumb pudding" model of the atom.

In 1911, Ernest Rutherford performed a classical experiment that threw the "plumb pudding model" into the dustbin of history. Using a piece of a radioactive element and a series of slits, Rutherford formed a well-collimated beam of alpha particles, and used this beam to probe the atomic structure of a metal foil. He knew that the alpha particles were thousands of times more massive than electrons, and would simply blast any electrons that they encountered out of the way. But what would happen when the positively-charged alpha particle interacted with protons within the metal atoms?

Rutherford directed a thin beam of alpha particles at a very thin film of metal. He monitored the position of the alpha particles after they passed through the metal by using a sheet of fluorescent material that glowed in the spot where an alpha particle hit.

Rutherford found that the vast majority (**nearly** all) of alpha particles travelled directly through the metal foil, without being deflected. Rutherford was a careful, meticulous and no doubt very patient experimenter. He found that a tiny fraction of the alpha particles was slightly deflected in their passage through the metal foil. And then, to his amazement, he found that an even tinier fraction (1 in 8,000 alpha particles) rebounded off the metal film! As Rutherford later related, this result was like firing a canon ball at a sheet of tissue paper, and finding some canon balls bouncing off the tissue paper.

A very short video describing Rutherford's experiment can be viewed at:
https://www.youtube.com/watch?v=_uEFKG122dA

To explain this result, Rutherford proposed that the protons within each metal atom were contained within an incredibly compact nucleus whose size is miniscule in comparison with the atom (that is, in comparison with the atomic orbitals occupied by its electrons). Most alpha particles would not pass very close to a nucleus, and would pass directly through the foil. A few would pass close to a positively-charged nucleus and be nudged sideways by the repulsion force. And a tiny minority would be moving directly towards a nucleus, and would rebound back in the direction from which they came.

Since the force of repulsion between a positively-charged alpha particle and a positively-charged nucleus is well-known from classical physics, and since the charges and masses of the alpha particle and the metal atom were known, it was possible to determine the maximum size of the nucleus of the metal atoms. Depending upon the particular element, the diameter of its atomic nucleus is 10,000-100,000 times less than the diameter of the atom (that is, its outer-shell electron orbitals). The miniscule size of the nucleus is even more striking when you consider that the volume of the largest nucleus is about *one-trillionth* that of its surrounding electron orbitals.

Imagine that we were located inside a hydrogen atom (this is entirely hypothetical). Its one electron would be located somewhere within a spherical-shaped atomic orbital. Let's say that the electron orbital has the same diameter as the dome of the Brisbane planetarium. The single proton comprising the nucleus of a hydrogen atom would be about 0.3 mm in diameter, smaller than the full stop at the end of this sentence. Thus, in any drawing depicting an atom, the nucleus is invariably shown grossly exaggerated in size (otherwise, anyone looking at the drawing couldn't see it).

Rutherford's experiment laid the groundwork for further developments in atomic physics. It was soon realised that the protons in the nucleus accounted for only about half the mass of an atom. To explain the "missing mass", Rutherford proposed that the nucleus contained particles with no electrical charge, with a mass about that of a proton. Being electrically uncharged, the proposed "neutron" would be more difficult to detect than electrons or protons, and many initial experiments failed to detect it. However, James Chadwick finally succeeded in detecting the neutron (for which he received the Nobel Prize in Physics in 1935) and even measured its mass. He bombarded deuterium with high-energy gamma rays (electromagnetic radiation that is identical to visible light, but with about a million times more energy), causing the deuterium nucleus to break apart, forming a lone proton and a neutron.

Since an atomic nucleus may contain many protons, you might be wondering why it is not simply blown apart by the huge repulsive forces acting between positively-charged protons so closely packed together. For many years, physicists pondered this question. The only plausible explanation was that an attractive "nuclear force" binds the protons and neutrons together in the nucleus. It was clear that this force must be very strong, but acts only over extremely short distances (about the diameter of a proton or neutron).

In small nuclei, the attractive nuclear force dominates over the electrical repulsive force, causing protons and neutrons to be tightly bound together. If we were to add more protons and neutrons to an atomic nucleus, the additional nuclear binding force would be greater than the repulsive force. Each additional proton or neutron contributes to the total "binding energy" of the nucleus. However, as we continue to add more protons and neutrons, we find that the nuclear binding energy increases at a diminishing rate. When we get to 56 protons (the element iron), the binding energy is a maximum. Beyond that, it is all "downhill" (for the nuclear binding energy).

For elements with nuclei larger than iron, the addition of more protons and neutrons actually detracts from its total nuclear binding energy. As more protons and neutrons are added, the nuclear force is not able to fully counter the increasing electrostatic repulsion between positively-charged protons, and the total nuclear binding energy decreases.

The largest nucleus found in nature is uranium, with 92 protons. Larger nuclei can exist, and indeed have been artificially produced using nuclear reactors or particle accelerators, but are unstable and undergo radioactive decay. Twenty-four synthetic elements have been produced, although some exist for only a fraction of a second. If any of these elements had been present when the Earth first formed, they would have long since decayed into other elements.

Uranium itself is unstable. A uranium nucleus will eventually eject an alpha particle or other radiation, but its "half-life" is billions of years. Much of the uranium that was embodied in the Earth (when it formed 4.5 billion years ago from a huge cloud of dust) is still present. That's a good thing for life on Earth, because radioactive decay of uranium-238 (as well as Thorium-232 and potassium-40) provides a heat source that keeps the mantle and core of the Earth at several thousand degrees. Heat released by radioactive decay keeps the Earth geologically active, and is responsible for the Earth's magnetic field, plate tectonics and the movement of continents across the Earth's surface. The amount of heat generated *inside* the Earth is estimated to be about 0.03% as much as the solar radiation absorbed by the Earth's surface.

Virtually all of the mass of an atom is due to the protons and neutrons in its nucleus. From the known atomic number of the various elements (giving the number of protons in the nucleus) and their atomic mass, we can work out the number of neutrons. For example, the element Carbon has an Atomic Number of 6 (with six protons), and an atomic mass almost exactly equal to 12 – and thus, has six neutrons. But, sometimes things are not so simple. Carbon has an atomic mass of 12.011, just slightly more than 12. It consists *almost* entirely of the isotope Carbon-12 (containing 6 neutrons), but also contains some atoms with 7 or even 8 neutrons in their nucleus (these are the isotopes Carbon-13 and Carbon-14 respectively).

Atoms with the same number of protons but different numbers of neutrons are called "isotopes". The isotopes of any particular element have virtually the same physical and chemical properties. So, a lump of carbon-14 would look the same as "normal" carbon-12 and would undergo the same chemical reactions, although it would be 16% heavier. However, different isotopes of an element can undergo very different nuclear reactions. Some isotopes are stable, while others decompose by emitting high-energy radiation. Radioactive isotopes are called, naturally enough, "radio-isotopes". As we have seen, some radio-isotopes eject an alpha particle (helium nucleus), while others eject high-energy electrons, neutrons or gamma rays (electromagnetic radiation similar to light, but with millions of times more energy).

No. of protons and neutrons in stable nuclei

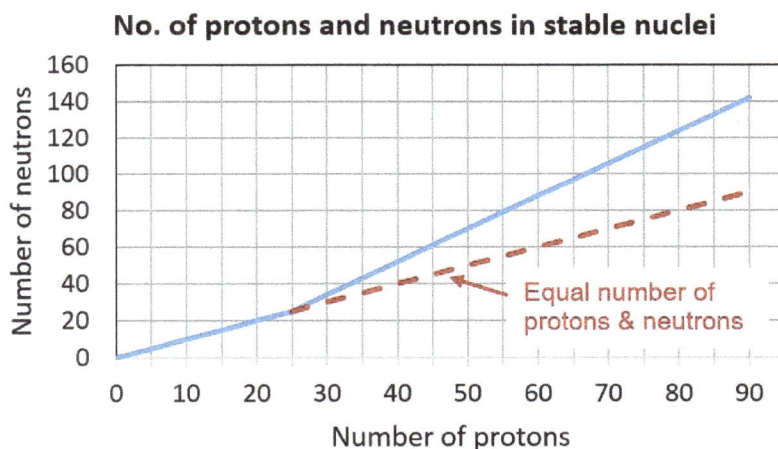

Normally, to be stable, small nuclei (containing up to 25 protons) have about equal numbers of protons and neutrons. Small nuclei having more (or fewer) neutrons than protons tend to be unstable and undergo radioactive decay.

Larger nuclei (containing more than 25 protons) must have more neutrons than protons to be stable. For example, the common isotope of uranium, Uranium-238, has 92 protons and 146 neutrons in its nucleus.

7. Atomic nuclei, and insights they provide into our past

We have seen that three types of particles - electrons, protons and neutrons – comprise all atoms in the universe.

Electrons reside in atomic orbitals, **well outside** the nucleus of atoms, and are responsible for the chemical properties of each element. Being negatively charged, they are attracted to the atom's positively-charged nucleus, but this binding is relatively weak. Relatively little energy is required to dislodge outer-shell electrons. Only 60-70 volts applied to fluorescent tubes can dislodge outer-shell electrons from numerous atoms of low-pressure gas inside the tube.

Protons and neutrons reside **inside** the nucleus, which comprises about one-trillionth the volume of the atom, but contains nearly all its mass (more than 99.9%). Protons and neutrons have very nearly the same mass, which is defined as one "atomic mass unit" (amu). The term "nucleon" is often used to refer to either a proton or a neutron. The nucleus is bound together by an incredibly strong "nuclear force", although this acts over very short distances, so that each proton or neutron only "feels" the nuclear force from adjacent nucleons. The energy that binds nucleons within the nucleus is millions of times greater than the energy that binds electrons within atoms, and is usually expressed in MeV (Millions of electron Volts).

As a result of the vastly different energies involved, nuclear processed are often considered in a completely different realm from chemical processes. In the past, nuclear reactions were considered the turf of nuclear physicists, and chemical reactions were the realm of chemists. Physicists were concerned with interactions of protons and neutrons **inside** the nucleus, and chemists were concerned with electrons residing in atomic orbitals **far outside** the nucleus. In recent decades, however, it has become clear that this distinction is not so black-and-white. In many areas of research, nuclear physicists and chemists are now working together on collaborative projects.

Let's come back to the three types of particles that are the basis of nuclear physics and chemistry. Properties of these particles are summarised in the table below.

	Charge	Mass	Symbol
Electron	-1	0.0005 amu	$_{-1}^{0}e$
Proton	+1	1 amu	$_{1}^{1}H$
Neutron	0	1 amu	$_{0}^{1}n$

I should mention that protons and neutrons are not the most fundamental atomic particles, but are composed of other exotic particles (which can be produced and studied by smashing protons together at enormous energies in particle accelerators like the Large Hadron Collider). These comprise an entire zoo of bizarre particles. Let me just mention two types of particles that are often produced by nuclear reactions:

- Neutrinos (and anti-neutrinos) are particularly noteworthy for their lack of interaction with matter. Trillions of neutrinos pass directly through the Earth each second, and hardly any are absorbed or interact in any way.
- Positrons have the same mass as an electron, and are identical to an electron in every way, except for their positive charge. Positrons are the "anti-matter" equivalent of an electron. When a positron is produced, it rarely gets very far before it strikes an electron, and the two particles literally "annihilate" each other. During the process, both particles disappear and their mass is converted entirely into the energy of two gamma rays (according to Einstein's formula $E = mc^2$). These gamma rays are electromagnetic radiation, like visible light, but with about one million times as much energy.

Protons and electrons are very stable, and can exist indefinitely (forever?). The Earth is constantly bombarded by "cosmic rays", a stream of protons ejected by the sun and probably other stars. Fortunately, the magnetic field of the Earth traps nearly all of these protons and prevents them from reaching the Earth's surface. However, unlike protons, neutrons are not stable outside the nucleus of an atom. A free neutron will eject a high-energy electron (beta particle) and form a proton with a "half-life" of 12 minutes. Bear in mind that 12 minutes is long enough for neutrons to travel perhaps thousands of kilometres.

The most abundant type of atom in the universe - and the simplest - is hydrogen, which contains one proton in its nucleus. We say that hydrogen has an "Atomic Number" of one - or in other words, has one proton in its nucleus, and must have one electron to give the atom an overall neutral charge. The vast majority of hydrogen atoms are "normal hydrogen", which contain no neutrons. With one proton and no neutrons, normal hydrogen has an "Atomic Mass" of one amu and is given the symbol shown here:

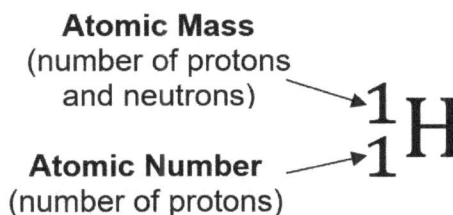

Atomic Mass
(number of protons and neutrons)

Atomic Number
(number of protons)

$$_1^1\text{H}$$

On Earth, we find that about one in 10,000 hydrogen atoms contains a neutron in its nucleus. This "isotope" of hydrogen is called "deuterium", and has virtually the same chemical properties as normal hydrogen, except for having twice the mass.

When I studied chemistry in high school, we were told that the various isotopes of an element have *virtually* identical chemical properties. The word "virtually" is an important caveat, as I learned later, because isotopes are not completely identical. The fact that a deuterium atom has twice the mass as normal hydrogen atoms does slightly affect its chemical and physical properties. For example, in water molecules containing deuterium atoms ("heavy water"), vibration of the hydrogen-oxygen bond occurs at lower frequency and with less energy. The reduced vibrational energy means that a molecule of heavy water in the liquid will be more easily trapped onto a crystal of frozen heavy water. Consequently, heavy water has a higher freezing point (3.8°C) than does normal water. Heavy water also boils at a slightly higher temperature than normal water (101.5°C). This makes it possible to separate heavy water from normal water which, as we shall see, is of fundamental importance for the Canadian nuclear power program.

Deuterium is still hydrogen – it has one proton in its nucleus, so its Atomic Number is still one, but its Atomic Mass is 2. Thus, the symbol for deuterium is $_1^2\text{H}$, although sometimes the letter "D" is used to symbolise deuterium as $_1^2\text{D}$.

As it turns out, a hydrogen atom can also have two neutrons in its nucleus, and this isotope is called "tritium". Tritium has the symbol $_1^3H$, and sometimes is symbolised with the letter T ($_1^3T$). However, the ratio of neutrons to protons in a tritium nucleus is too high for it to remain stable, and it undergoes radioactive decay. The half-life of tritium is 12.3 years so, while tritium can be produced in nuclear reactors or particle accelerators, it does not exist naturally. Any tritium that might have existed when the Earth formed would have decayed long, long ago. When a tritium nucleus decays, one of its neutrons ejects a high energy electron (Beta particle), leaving an additional proton in the nucleus. The end product is an atom of helium-3 (containing two protons and one neutron), as well as a Beta particle (high-energy electron) and an anti-neutrino.

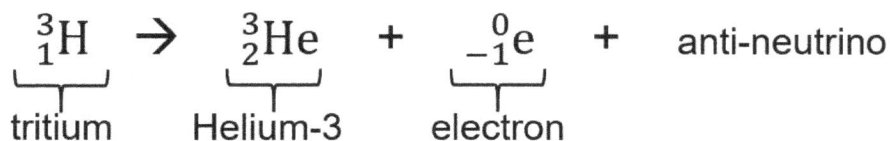

$$\underbrace{_1^3H}_{tritium} \rightarrow \underbrace{_2^3He}_{Helium\text{-}3} + \underbrace{_{-1}^{0}e}_{electron} + \text{anti-neutrino}$$

Note that, when we write nuclear reactions such as this, the sum of the Atomic Numbers (the number of positive charges in the nuclei) on the left side of the equation is the same as on the right side of the equation (1 = 2 -1). Similarly, the sum of the Atomic Masses on the left side of the equation is equal to the sum on the right side of the equation (3 = 3 + 0).

Isotopes that contain too many neutrons to be stable tend to emit a high-energy electron. One important example is the isotope potassium-40, which is used to date the formation of rocks. Potassium-40 has a half-life of 1.25 billion years, so some of this isotope that was present when the Earth formed (4.5 billion years ago) is still present in the Earth's mantle and crust. The main mechanism for the radioactive decay of potassium-40 is the emission of a Beta particle (high-energy electron). Once again, a neutron decays into a proton (which remains in the nucleus) and an electron. Energy released by the process is imparted as kinetic energy to the electron. During the process, the number of protons in the nucleus increases by one, yielding a nucleus of calcium-40.

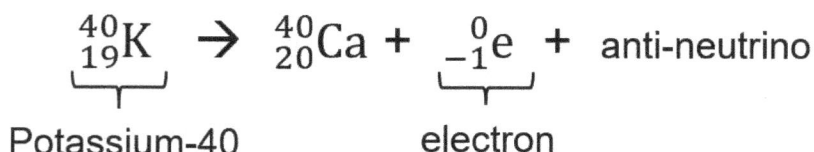

$$\underbrace{_{19}^{40}K}_{Potassium\text{-}40} \rightarrow _{20}^{40}Ca + \underbrace{_{-1}^{0}e}_{electron} + \text{anti-neutrino}$$

About 90% of potassium-40 atoms decay in this way. However, about 10% of the potassium-40 atoms decay via a different mechanism called "electron capture". In this case, an electron in an inner-shell orbital near the nucleus is captured by the nucleus, converting a proton into a neutron. The reaction is shown below:

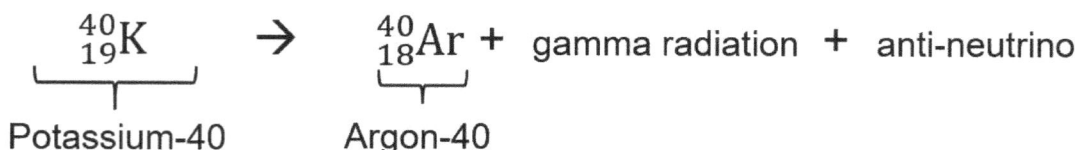

$$\underbrace{_{19}^{40}K}_{Potassium\text{-}40} \rightarrow \underbrace{_{18}^{40}Ar}_{Argon\text{-}40} + \text{gamma radiation} + \text{anti-neutrino}$$

In this case, when we add the Atomic Numbers on the left side of the equation, we get a **different** sum than on the right side of the equation (19 does **not** equal 18)! "Electron capture" doesn't follow the usual rule because the Atomic Number of potassium-40 only includes the charged particles (protons) **inside** the nucleus, while the argon-40 nucleus includes an electron which was captured from **outside** the nucleus.

The really interesting thing about this reaction is that it produces argon, an element which is completely chemically inert. Argon doesn't react chemically with anything, and so, when minerals solidify from molten magma, they never contain argon bound within the crystal structure. Consequently, if we find any argon atoms trapped within the crystal structure of a mineral, they must have been formed by radioactive decay of potassium-40 **after the mineral crystal formed**. By measuring the concentration of argon and potassium isotopes within a rock, and knowing the half-life for the decay of potassium-40, it is possible to determine how long ago its minerals crystallised from molten magma.

As a result of potassium/argon-40 isotope analysis, and uranium-lead isotope analysis, we can determine when rocks formed in continental and ocean crust. This technique has provided additional evidence for plate tectonics and the movement of continents across the Earth's surface. Near mid-ocean ridges, where partially-molten rocks rise from the mantle to form new ocean floor, the rocks are quite young and contain hardly any argon. Further away from mid-ocean ridges, rocks underlying the seafloor are older, reaching a maximum age of 200 million years where "oceanic plates" are subducted beneath "continental plates". Continental rocks can be much older. Rocks have been found in Western Australia containing zircon mineral crystals (probably eroded from earlier rocks) formed 4.4 billion years ago. These are the oldest minerals that have been found anywhere on Earth, and probably crystallised soon after the Earth formed and its crust cooled.

Another isotopic technique, using carbon-14, is used to date fossils from human settlements during the past 50,000 years (including Neanderthals, which became extinct about 30,000 years ago, as well as Denisovans and Homo floresiensis in Indonesia).

Carbon is an essential element in plants, animals and all known living things. It forms the chemical backbone of proteins, carbohydrates, DNA and RNA. Carbon is constantly circulating between the atmosphere and the biosphere (the bodies of all living things). Plants absorb carbon dioxide from the air by photosynthesis, and the carbon is absorbed when animals eat plants (or when carnivores eat other animals that eat plants). When humans or other animals metabolise food (respiration), carbon is returned to the atmosphere as carbon dioxide.

In nature, carbon consists **almost** entirely of carbon-12, containing six protons and six neutrons in its nucleus. However, the Earth is constantly bombarded by high-energy protons from the sun, and when these enter the upper atmosphere, they smash into oxygen and nitrogen atoms to produce a shower of neutrons and other sub-atomic particles. When a neutron hits the nucleus of a nitrogen atom, it becomes bound to the nucleus and causes a proton to be ejected. The product of this nuclear exchange is carbon-14. This process has almost certainly been occurring for many millions (probably billions) of years. Here is the reaction:

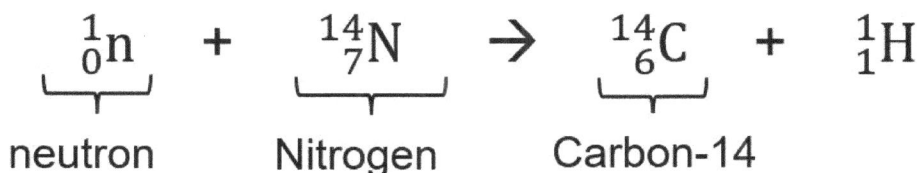

$$\underbrace{^{1}_{0}n}_{\text{neutron}} + \underbrace{^{14}_{7}N}_{\text{Nitrogen}} \rightarrow \underbrace{^{14}_{6}C}_{\text{Carbon-14}} + {}^{1}_{1}H$$

Carbon-14 is constantly being formed in the upper atmosphere, from where it circulates and mixes with the rest of the atmosphere and is absorbed by living plants. While carbon-14 is constantly being formed, it is also being destroyed at the same rate by radioactive decay. As a result, the concentration of carbon-14 remains constant at a level (about one atom per trillion carbon atoms) where it is formed at the same rate as it is destroyed. Carbon-14 decays back to nitrogen by emitting a Beta particle (high-energy electron). Its half life is 5,730 years.

$$^{14}_{6}C \quad \rightarrow \quad ^{14}_{7}N \ + \ ^{0}_{-1}e \ + \ \text{anti-neutrino}$$

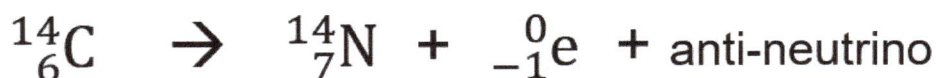

So, all living things contain carbon-14 within their tissues at a concentration of about one part-per-trillion. However, once creatures die, they no longer exchange carbon with the atmosphere to maintain the concentration of carbon-14 in their tissues. Plant remains from, say, 5,700 years ago contain carbon-14 at half the level of living plants, and remains from 11,400 years ago would have one-quarter the concentration of carbon-14. Fossil fuels like coal, formed from the remains of plants and algae that lived many millions of years ago, contain virtually no carbon-14.

Consequently, by measuring the level of carbon-14, we can accurately assess the age of plant and animal remains. Often, animal or human remains that are thousands of years old have long since decayed, leaving only bones and teeth. However, ancient human societies often lived and were buried near their campfires, and carbon in the ash can be dated by carbon-14 analysis. Charcoal was often used in cave paintings, so we can often know quite accurately when prehistoric artists adorned the walls of their caves.

Dating of ancient remains by carbon-14 analysis is very accurate, but must be done very carefully to ensure that ash or charcoal originates from the same period as human remains associated with it. The results must be calibrated to take into account the global impact of mankind in recent times. In particular, large-scale burning of fossil fuels since the industrial revolution has released "old" carbon, containing virtually no carbon-14, back into the atmosphere. Testing of nuclear weapons in the atmosphere by the United States, Soviet Union, England and France in the 1950s and 1960s has significantly increased the level of carbon-14 in the air, and this will need to be taken into account if archaeologists discover our remains thousands of years from now.

Carbon-14 analysis provides a method to monitor carbon dioxide emissions from coal-fired and gas-fired power stations. In the past, governments relied on power station operators to provide accurate information on the type and amount of coal that was burned, enabling the amount of carbon dioxide emitted to the atmosphere to be determined. However, governments need to be able to evaluate carbon dioxide emissions independently to obtain objective and reliable greenhouse emissions, One clever method collects samples of grass or other plants growing downwind of a power station exhaust stack. By extracting carbon dioxide from the air, and using the carbon to make sugars and carbohydrates, plants provide a record of average carbon-14 content of the air that they "breathe". Since the carbon content of coal and natural gas contains virtually no carbon-14, exhaust emissions of power stations contain a far smaller concentration of carbon-14 than the atmosphere. By measuring the carbon-14 content of grass growing downstream of a power station, researchers have been able to calculate carbon dioxide emissions from the power station within 10% accuracy [Reference 1].

References
1. Independent evaluation of point source fossil fuel CO_2 emissions to better than 10%, Proceedings of the National Academy of Sciences (US) 2016. http://www.pnas.org/content/early/2016/08/24/1602824113

8. Radioisotope power generators

Most satellites orbiting around the Earth rely on solar panels to provide electrical power they need to operate their instruments and maintain communications with ground-based controllers. However, solar panels are not suitable for missions into deep space, further from the sun. For example, the planet Neptune is 30 times further from the sun as is the Earth. At this distance, the intensity of solar radiation is $(1/30)^2$, or 900 times less than solar radiation reaching the Earth. From Neptune, the sun would be only slightly brighter than other stars in the sky, and solar panels would be virtually useless for producing electrical power.

With the limits of current rocket technology, any space mission to the outer parts of the solar system takes years, and must have electrical power to operate instruments, transmit photographs and measurements, and receive instructions from ground controllers. Batteries and fuel cells do not have the capacity to provide even limited amounts of power for such extended periods of time.

Power for spacecraft missions to deep space is provided by Radioisotope Thermal Generators (RTGs). These utilise an isotope that undergoes radioactive decay over a half-life that roughly matches the duration of the mission. Radiation emitted by these isotopes (alpha particles, beta particles or gamma radiation) is absorbed and heats the surrounding material. These materials are not only "hot" in a radioactive sense, they are literally "hot" in temperature. The rate of radioactive decay depends only on the half-life of the isotope (and not on its temperature or environment), so very high temperatures can be produced (typically, several hundred degrees C).

Radioisotope generators utilise high temperatures produced by their radioactive heat source to produce electricity in a thermoelectric panel. Thermoelectric panels are solid-state devices that are very similar to solar panels – but instead of converting light into electricity, they generate electricity from the temperature difference across a semiconductor junction. They have no moving parts and can operate extremely reliably for many years, although their heat-to-electricity conversion efficiency is poor. Like all heat engines, only a portion of the heat can be converted into mechanical power or electricity, and remaining heat must be dissipated into deep space.

In a Radioisotope Thermal Generator, thermoelectric panels are wrapped around a cylindrical core which is heated by radioactive decay. The inner surface of the thermoelectric panel is heated to high temperature, and the outer surface is cooled by fins, whose large surface area radiates heat as infrared radiation (since there is no air in space to carry away the waste heat, it must be emitted as radiation).

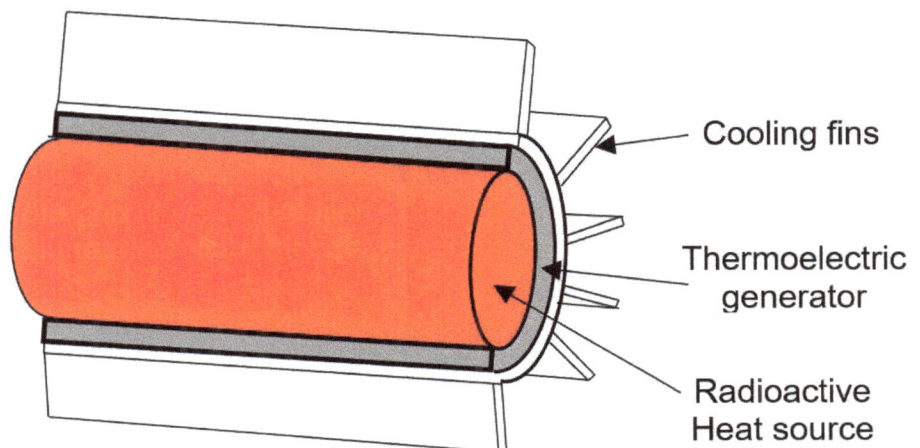

Cooling fins

Thermoelectric generator

Radioactive Heat source

Cross-section of a typical Radioisotope Thermal Generator

Radioisotope Thermal Generators should not be confused with conventional nuclear power stations. RTGs operate through an entirely different nuclear process, are small and compact, and produce relatively small amounts of electrical power (a few hundred watts - about one millionth as much as a conventional coal or nuclear power station).

Radioisotope thermal generators have been used in nearly all of the historic space missions: Pioneer 10 and 11, Voyager 1 and 2, Galileo, Ulysses, Cassini, New Horizons, the Viking landers, and to power instruments left on the moon by the Apollo astronauts. First developed in the 1950s and 1960s, they were enthusiastically expected to be used everywhere, despite the potential safety risks posed by radioactive material. About 1,000 Radioisotope Thermal Generator units were installed in remote areas of the Soviet Union. All are now well past their 10 year design lifetime, and apparently, some lie derelict and forgotten or vandalised. About 250 miniature plutonium-powered generators were manufactured to provide power to cardiac pacemakers. Remarkably, 22 are reported to still be in operation 25 years later.

The launch of spacecraft containing kilograms of highly radioactive and toxic material presents significant environmental and health risks. Early in the space program, in the 1960s, several launch failures resulted in release of radioactivity. Perhaps to downplay the risk posed by launching radioactive material into space, the term "SNAP generator" began to be used (an acronym for "**S**ystems for **N**uclear **A**uxiliary **P**ower").

Improved designs of radioisotope generators were made very rugged, and designed to withstand an explosion on launch or uncontrolled re-entry through the atmosphere. These RTGs were put to the test during the ill-fated Apollo 13 moon mission The SNAP generator on-board the Apollo 13 lunar lander was supposed to have remained on the moon. Instead, after a catastrophic explosion in an oxygen tank on-board the service module, the Apollo 13 lander served as a "lifeboat" for the crew to return to Earth. As the lander, service module and command module approached the Earth, the lander was evacuated by the crew, jettisoned from the Command Module, and burned up as it re-entered Earth's atmosphere at more than 40,000 kilometres/hour. Its SNAP generator survived intact and is now lying on the bottom of the Pacific Ocean.

Many isotopes are radioactive, but most are not suitable for application in SNAP generators. Since radioactive emissions cannot be turned on and off, and the rate of radioactive decay is fixed, the isotope must have a half-life that matches the expected life of the mission for which power will be required. Ideally, the radioisotope should have a high "power density", so that only a small amount of the material is needed to produce the required amount of heat and electrical power. However, there is a general trade-off between half-life and power density. Radioisotopes that have long half lives decay at a slow rate, and have low power density. Also, the radioisotopes should give off a type of radiation that is absorbed within a small thickness of material, avoiding the need for massive amounts of shielding. For this reason, radioisotopes that emit gamma radiation are generally not suitable. Nearly all SNAP generators use radioisotopes that emit alpha particles (helium nuclei moving at high velocity). This is the easiest type of radiation to absorb, and alpha particles are emitted with about ten times as much energy as Beta particles.

The most common radioisotope used in the US space program is plutonium-238. This isotope was first synthesized in 1941 by the US chemist Glenn Seaborg, who became one of the most prominent and famous chemists of the 20th century. Around 1981, when I was editing an international chemistry magazine, I was gobsmacked to receive a letter from Glenn Seaborg (although the content of the letter was completely mundane).

Left: Inspection of SNAP generators on the Cassini spacecraft prior to launch to Saturn (source NASA).

Right: the New Horizons spacecraft, with two SNAP generators extending on a side arm, recently encountered Pluto (source: PD-USGov-NASA). Each generator unit contains 7.8 kilograms of Pu-238 and produces 300 watts of electrical power.

Each nucleus of plutonium-238 emits an alpha particle, leaving a nucleus of uranium-234. The alpha particle has an energy of 5.6 million electron volts, which is converted to heat when the radiation is absorbed.

$$^{238}_{94}\text{Pu} \;\rightarrow\; ^{234}_{92}\text{U} \;+\; ^{4}_{2}\text{He} \text{ (alpha particle with 5.6 MeV)}$$

Note that plutonium-238 is a different isotope than plutonium-239, which is produced by nuclear reactors (and used as a nuclear fuel in reactors and in nuclear bombs). Plutonium-238 is 300 times more radioactive than plutonium-239. It is produced by irradiating either neptunium-237 or Americurium-238 with neutrons in a nuclear reactor. The US stopped producing plutonium-238 in 1988, as US nuclear agencies obviously felt they had enough. Since then, the US has purchased plutonium-238 from Russia, and is evidently planning to resume production.

Here are key characteristics of radioisotopes that are most commonly used in SNAP generators. One kilogram of each of these radioisotopes releases about the same total energy over its half-life, but their half-lives are very different. Heat production reduces at an exponential rate, so the choice of radioisotope depends on the duration of the mission. Missions to the outer solar system can take 5 or ten years, or even longer, so plutonium-238 provides a good match.

Radioisotope	Half-life	Power density	
Plutonium-238	87.7 years	540 watts/kilogram	Lowest shielding requirement.
Polonium-210	138 days	140,000 watts/kilogram	Very high power density, but short half-life
Americurium-241	432 years	120 watts/kilogram	Lower power density, but very long half-life.

Since the heat output of plutonium-238 declines over time, as it does for all radioisotopes, the amount of plutonium-238 must be sufficient to provide the power required at the end of a mission. Ultimately, after all atoms of plutonium-238 have decayed, each kilogram of plutonium-238 will have produced 2,200 billion joules of heat. Most SNAP generators convert less than 10% of the heat energy into electricity, so the total electrical energy output from one kilogram of plutonium-238 is about 200,000 megajoules. This is about 25,000 times the electrical energy output produced from one kilogram of liquid hydrogen and liquid oxygen in a fuel cell.

9. Nuclear fusion: Energy source for the sun and stars

Our star, the sun, produces prodigious amounts of energy. Only a miniscule fraction of solar radiation emitted by the sun strikes the Earth, yet this accounts for virtually all the energy that drives the wind, ocean currents, weather, photosynthesis and the living creatures that inhabit the biosphere. Yet, until the 1920s, scientists had no idea of the energy source that powers the sun. Various calculations had been done to determine how long the sun could operate using known energy sources (like the burning of coal), but these could barely account for a lifetime of even a million years. But, by the 1800s, it was already becoming obvious (from geologic and fossil evidence) that the Earth must be much older than that.

With the advances in nuclear physics in the early 20th century, it quickly became apparent that the only plausible mechanism to account for the huge energy release by the sun is nuclear fusion.

The fusion of small nuclei (like hydrogen) into larger nuclei is driven by the nuclear binding force. The binding energy of a nucleus of helium, containing two protons and two neutrons, is greater than that of four separate hydrogen nuclei, so energy is released when four hydrogen nuclei "fuse" into a helium nucleus.

With helium having four nucleons (two protons and two neutrons), one might expect that a helium atom would have almost exactly the same mass as four hydrogen atoms. However, in the 1920s, precise measurements showed that an atom of helium has about 0.7% less mass than four hydrogen atoms (4.0313 atomic mass units for helium versus 4.0026 amu for four hydrogen atoms). Arthur Eddington surmised that if four hydrogen nuclei were "fused" into a helium nucleus, the "missing mass" would be converted into energy according to Einstein's equation $E=mc^2$. The stupendous energy released by fusion of four hydrogen atoms into helium would be enough to generate temperatures of **billions** of degrees. This huge release of energy could readily account for the operation of stars, and this was verified by experiments (initially, in developing thermonuclear, or "hydrogen" bombs, and in the decades since, in experimental nuclear fusion reactors).

To get hydrogen nuclei to "fuse" is not easy - not in the laboratory. The nuclear force that binds protons and neutrons together in a nucleus operate at extremely short distances (about the diameter of a proton), so hydrogen nuclei must be brought very closely together for fusion to occur. However, hydrogen nuclei (protons) have a positive charge, and therefore, repel one another. To overcome this repulsion, and to get protons close enough to fuse, they must collide at extremely high velocity. Such velocities can be achieved in high-energy particle accelerators, or by heating hydrogen to a temperature above 10 million degrees. Such high temperatures occur near the centre of stars.

The sun, like other stars, is a huge spherical ball of hot gas, consisting primarily of hydrogen, with some helium and trace amounts of other elements. With a surface temperature of about 6,000 degrees Kelvin, it radiates electromagnetic radiation – visible light (a mixture of all colours that we can detect with our eyes), infrared radiation and ultraviolet.

Being so hot, with atoms smashing into each other at high velocity, electrons in the atomic orbitals of hydrogen (and other gases) are ejected from atoms. This leaves a "plasma", a hot soup of protons and electrons moving in all directions. This plasma acts like a gas – it is compressible and exhibits an outwards pressure (force per unit area). In fact, the plasma in the sun acts like an "ideal gas", for which its outwards pressure varies directly with the number of particles per unit volume, and with its absolute temperature.

Imagine the plasma contained within a hollow spherical shell inside the sun. This plasma has mass, which is pulled towards the centre of the sun by gravitational attraction. An outwards force is required to counter this gravitational attraction, and to prevent the plasma shell from collapsing into the centre of the sun.

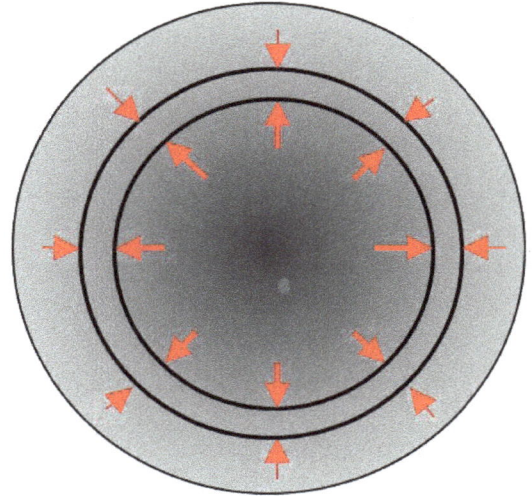

The outwards force is provided by pressure at the inside surface of the shell being greater than the pressure on its outside surface. Consequently, for the sun to resist collapsing under its own weight, the pressure must increase towards the centre of the sun. This increasing pressure arises mostly from increasing temperature towards the core of the sun (as well as from increasing density of the plasma).

Bear in mind, Earthlings, that the sun has 300,000 times as much mass as the Earth, so gravitational forces are much greater in the sun than on Earth [Ref 1]. Of course, the sun is also much larger than the Earth. Overall, at the "surface" of the sun, gravitational forces are twenty seven times what we experience at the surface of the Earth. In other words, your weight on the sun would be twice the weight of a small car on Earth. If the sun had a solid surface (which it doesn't), you would be squashed into a blob, if you didn't vaporise first.

Below the surface of the sun, its huge mass and gravity causes the pressure to rise to unimaginable, crushingly high values. Consequently, to prevent the sun from collapsing under its own weight, the temperature of the plasma rises steadily with increasing depth. Near the centre of the sun, within its "core" region, the pressure, temperature and density reach their maximum values. Using fairly straightforward physics and maths, it is possible to estimate that the pressure reaches *250 billion atmospheres*, the temperature reaches about *15 million degrees*, and the density is 150 times that of liquid water.

The core temperature of 15 million degrees is high enough to maintain the nuclear fusion reaction which provides energy for the sun. This is not a coincidence. *If* the temperature were *not* hot enough to enable nuclear fusion to occur, the sun would begin to cool and contract under its own weight, causing the temperature at the core to increase until nuclear fusion started.

Thus, the temperature at the centre of centre of the sun is *just enough* to maintain nuclear fusion at the rate needed to maintain its core temperature and prevent the star from collapsing. The nuclear fusion reaction provides just enough energy to offset losses of energy by radiation (visible, infrared and ultraviolet) from its surface layer.

Like all main-sequence stars, the sun is self-regulating:

- If the core of the sun cooled below the temperature required for nuclear fusion, the star would begin to collapse, and this collapse would increase the temperature of the core until nuclear fusion resumed.

- If the core of the sun became too hot, the nuclear fusion reaction would occur faster and produce more heat than could escape. This would initially cause the pressure inside the core to rise, pushing out surrounding layers of plasma. As the star swelled, its density and gravitational attraction would reduce, and this would then reduce the temperature and pressure in the core and slow the rate of nuclear fusion.

This self-regulating mechanism causes our sun (and similar stars) to provide a stable output of light and heat. Such a stable solar output was essential for life to arise and evolve on Earth, and is still essential for life to survive on our planet.

The conditions inside a star, and the rate at which energy is produced by nuclear fusion, depend dramatically upon the mass of the star. Stars that are smaller than our sun have a lower temperature in their core, which cause the nuclear fusion to occur at a much reduced rate. Because they are "burning" hydrogen more slowly, low-mass stars have a greatly increased life-span. For example, a star that has half the mass of the sun would emit only about 3% as much light. Its surface temperature would be about 3,800°K, much cooler than the surface of our sun. Because such small stars undertake nuclear fusion at a very slow rate, the hydrogen within its core would last for about 200 billion years (fifteen times the present age of the universe) before it is all converted to helium.

There is a lower limit to how small a star can be. Once the mass is less than about 8% of the mass of the sun, the temperature at its core is too low to sustain nuclear fusion.

Conversely, stars that are larger than the Earth have higher core temperatures and fuse hydrogen to helium at a furious rate. Massive stars "live fast and die young". The amount of energy produced by nuclear fusion, and radiated from a star's surface, varies roughly with the fourth power of its mass. *A star that is three times the mass of our sun would emit* about 3 x 3 x 3 x 3 = *sixty times as much radiation*. Its surface would have a temperature of 11,000°K, and would emit light that appears blue in colour and contains a high proportion of ultraviolet radiation. Within a mere 200 million years, such a star would fuse all of the hydrogen in its core into helium. At that point, it would "leave the main sequence" and start to fuse helium into heavier elements (lithium, beryllium, boron, carbon and oxygen). However, these nuclear reactions liberate far less energy than fusion of hydrogen. Once a star of this size leaves the main sequence, it has a relatively short life before it eventually fades away as a "brown dwarf".

More massive stars end their lives in a violent supernova explosion. For a brief time, the star outshines billions of other stars in its galaxy. Supernova explosions produce an incredibly intense blast of neutrons (as well as gamma rays and other particles), which stream outwards at nearly the speed of light and strike atomic nuclei blown out from the explosion. These neutrons become bound to atomic nuclei which they hit. Some of the added neutrons decompose, forming additional protons in the nucleus and ejecting a high-energy electron (Beta particle).

Repeated bombardment by high-energy neutrons forms very large nuclei. Elements that are more massive than iron, with 26 protons in its nucleus, have less binding energy than smaller nuclei. Such massive nuclei are only formed because the kinetic energy of impacting neutrons provides the energy required for these reactions. This is how heavy elements like gold and uranium are produced, and spewed across galaxies, during the final dying moments of massive stars undergoing supernova explosions.

Evidence that the Earth has periodically been showered by remnants of supernova explosions was reported in 2016 [Reference 2]. Traces of the radioisotope iron-60 were discovered in sediments deposited about 2 million and 8 million years ago. Iron-60 has a half-life of 2.6 million years and is not normally found on Earth. Each of these bands of Iron-60 is thought to result from a supernova explosion occurring at a distance of 300 light years from Earth. Incidentally, any supernova occurring within about 30 light-years from Earth would probably cause a mass extinction of life on this planet.

Here is a table comparing the luminosity (light energy emitted), surface temperature and lifetime of stars, depending upon their mass (relative to our sun). You can see that the mass of a star exerts an enormous effect on how a star lives and dies. It is almost as if small stars and large stars are different "species".

Star mass (relative to the sun)	Luminosity (relative to the sun)	Surface temperature	Lifetime on main sequence
0.5	0.03	3,800°K	200 billion years
0.75	0.3	5,000°K	30 billion years
1.0	1	6,000°K	10 billion years
3	60	11,000°K	200 million years
5	600	17,000°K	70 million years
15	17,000	28,000°K	10 million years
25	80,000	35,000°K	7 million years

Source: An introduction to the sun and stars. Chap 6: The main sequence life of stars, S.F. Green and M.H. James, 2015, Available in Qld State Library 523.8 2015.

As it turns out, most stars are less massive than the sun. Such low-mass stars are much, much dimmer and more difficult to see. Their surfaces are relatively cool, so they emit mainly infrared radiation, rather than visible light. Low-mass stars have recently attracted great interest in the quest for planets that could harbour life. Although their light output is relatively dim, planets orbiting close to these stars could be within the temperature range for liquid water. The light output of such stars remains constant for many billions of years, providing stable conditions for living organisms to appear and evolve.

The fusion of hydrogen into helium occurs in several steps. It might be worthwhile having a good look at this reaction, as this is probably the single most important process occurring in the universe.

Here is the sequence of reactions occurring in our sun (and stars of comparable mass):

(1) Firstly, two hydrogen nuclei (protons) collide, forming a deuterium nucleus, a positron and a neutrino (μ):

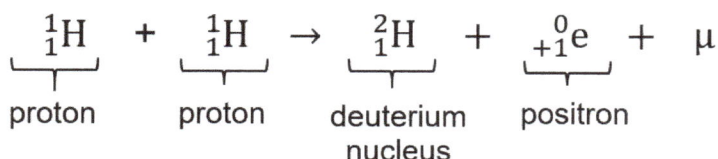

$$ \underbrace{{}_{1}^{1}\text{H}}_{\text{proton}} + \underbrace{{}_{1}^{1}\text{H}}_{\text{proton}} \rightarrow \underbrace{{}_{1}^{2}\text{H}}_{\substack{\text{deuterium} \\ \text{nucleus}}} + \underbrace{{}_{+1}^{0}\text{e}}_{\text{positron}} + \mu $$

This initial reaction is the "rate-limiting step" for the overall process. One of the protons must decay into a positron and a neutron during the momentary instant that the two protons are in close proximity. This is a relatively rare occurrence. A proton in the sun's core might smash into its colleagues for a billion years before finally colliding with sufficient energy to produce a deuterium nucleus. The rate of this reaction is extremely sensitive to temperature. Even a "slight" increase (relatively speaking) of a few million degrees, increases the reaction rate a thousand times.

(2) Once deuterium forms, it can then react relatively quickly with another proton to form a helium-3 nucleus and a gamma ray (γ).

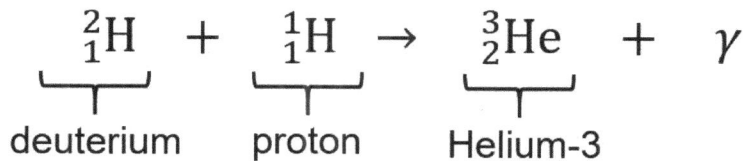

$$\underbrace{{}^{2}_{1}H}_{\text{deuterium}} + \underbrace{{}^{1}_{1}H}_{\text{proton}} \rightarrow \underbrace{{}^{3}_{2}He}_{\text{Helium-3}} + \gamma$$

(3) There are several ways that helium-three can react further. In our sun, most helium-3 nuclei react with another helium-3 nucleus to form helium-4, ejecting two protons in the process.

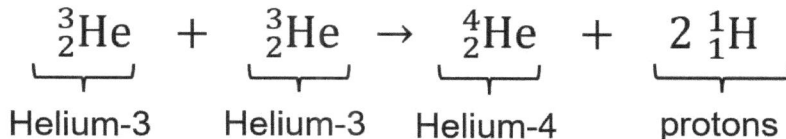

$$\underbrace{{}^{3}_{2}He}_{\text{Helium-3}} + \underbrace{{}^{3}_{2}He}_{\text{Helium-3}} \rightarrow \underbrace{{}^{4}_{2}He}_{\text{Helium-4}} + \underbrace{2\,{}^{1}_{1}H}_{\text{protons}}$$

Note that the positrons formed in step 1 do not last very long. Almost immediately, each positron will encounter an electron, and the two particles annihilate each other. The mass of the positron and electron are converted entirely into the energy of two gamma ray photons.

The overall reaction is thus:

$$4\,{}^{1}_{1}H + 2\,{}^{0}_{-1}e \rightarrow {}^{4}_{2}He + 6\gamma + 2\mu$$

Four hydrogen nuclei (protons) and two electrons combine to form a helium-4 nucleus, six gamma ray photons and two neutrinos. This reaction liberates 26 million electron volts. To put that into perspective, the fusion of one kilogram of hydrogen produces about the same energy as the combustion of 24 million kilograms of coal. About 2% of the energy released by the fusion reaction is carried by the neutrinos, which travel unhindered through the sun and escape into space. The remaining 98% of the energy produced by the fusion reaction is carried by gamma ray photons and kinetic energy of helium and hydrogen nuclei.

Energy released by the fusion reaction must eventually find its way outwards from the core of the sun, reach the surface, and escape as visible, infrared and ultraviolet radiation. To reach the sun's surface, a gamma ray photon must pass through a thick "soup" of free electrons and protons

Gamma ray photons produced by the fusion reaction are extremely energetic: each photon has an energy of about a million electron volts (compared to visible light photons with energies of 2-3 electron volts). Such high-energy photons carry considerable momentum, as well as energy. For visible light, the momentum is so small as to be barely measurable. Not so for gamma rays. In 1922, Arthur Holly Compton showed that when high-energy x-rays or gamma rays collide with electrons or other particles, they are scattered like one billiard ball hitting another. During such a collision, a gamma ray photon transfers some of its momentum and energy to the electron (depending upon the angle that the gamma ray is deflected). This process was later named the "Compton Effect".

Imagine a gamma ray photon hitting a stationary electron and being scattered at some angle θ. The scattering angle θ may vary from 0° (undeflected from its original direction) to 180° (rebounding back in direction from which it came). Depending on the angle at which the gamma ray is scattered, the gamma ray photon imparts some of its energy and momentum to the electron. The scattered gamma ray has lower energy and frequency than the incident gamma ray photon.

45

Based on the conservation of momentum and conservation of energy, it is possible to derive an equation for the amount of energy ΔE lost by a gamma ray photon during a collision with an electron.

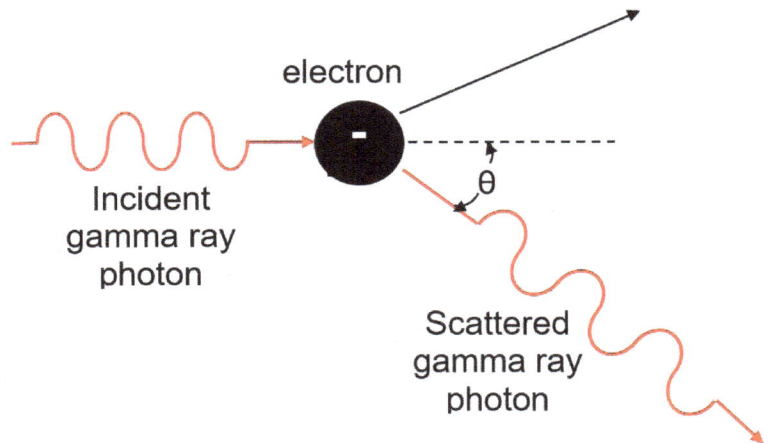

Incident gamma ray photon

electron

Scattered gamma ray photon

$$\Delta E = 2\,\frac{E^2}{m_e c^2}\,(1 - \cos\theta)$$

Where **E** is the original energy of the gamma ray photon
m_e is the mass of the electron
c is the speed of light (3×10^8 metres/second)
θ is the angle that the gamma ray photon is scattered

The gamma ray photon loses the greatest amount of energy when the scattering angle θ is 180°, that is, when the gamma ray is scattered back in the direction from which it came.

Because the initial energy **E** of a gamma ray photon is huge, it loses a significant fraction of its original energy in each collision.

The plasma deep inside the sun is very dense (as we have seen, the density at the core is about 150 times the density of water), so gamma ray photons travel only a tiny distance before they collide with an electron. Gamma ray photons undergo perhaps billions of collisions each second. If we could see the collisions between a gamma ray photon and electrons in the "plasma soup", it might look like balls in the pinball machines of the 1950s, but speeded up a billion-fold.

The escape path of the gamma ray photons is long and tortuous. The distance from the sun's core to its surface is about 600,000 kilometres, which would only take 2 seconds to traverse if the photons travelled in a straight path. But the journey for the gamma rays is convoluted and much, much longer. Each time a photon is scattered, it is about as likely to be reflected back in the direction from which it came, as to continue forward. Even though the photons are travelling at the speed of light, it takes an estimated 50,000 years for a photon to eventually reach the surface.

Along their outward journey, the photons are constantly transferring energy **to** electrons, and sometimes gaining energy **from** electrons, during countless collisions. Bear in mind that the Compton Effect is reversible, so the exchange of energy between photons and electrons can occur in either direction. Initially, gamma ray photons have much more energy and momentum than the electrons they encounter, and the photons lose energy and momentum during each collision. However, after countless collisions, a photon of radiation has about the same energy as the electrons it encounters. Then, the photon is nearly as likely to *gain energy* when it rebounds off a fast-moving electron than to transfer energy to a slower electron. In terms that scientists would use, we say that the photons reach "thermal equilibrium" and have the same energy distribution as the electrons and other particles that they encounter [Note 1].

At the same time, hot electrons, protons and helium nuclei within the plasma are constantly **producing new photons** of "blackbody radiation", whose energy reflects the temperature of the plasma [Note 2]. These new photons join the melee of photon-electron collisions, and they too migrate outwards and transfer energy to cooler plasma near the surface.

So, as gamma rays gradually migrate outwards, they transfer energy to electrons, protons and (indirectly) to new photons produced in the surrounding plasma. The "gamma rays" become "hard x-rays", then "soft x-rays", and then ultraviolet radiation. By the time they reach the surface of the sun, the photons have become the visible, infrared and ultraviolet light that is characteristic of blackbody radiation emitted by a surface at 6,000 degrees (the temperature at the surface of the sun). Each photon of light that escapes the sun's surface has about one-millionth the energy of the gamma rays produced by nuclear fusion in the sun's core.

References

1. For basic facts about the sun, see: http://nssdc.gsfc.nasa.gov/planetary/factsheet/sunfact.html
2. "Traces of exploding stars litter the Earth and Moon", Cosmos magazine, Issue 69, 2016, page 12.

Notes

1. Electrons are not the only particles that scatter photons. Gamma ray photons are also scattered by collisions with protons and helium nuclei, but note from Equation (1) that the energy loss during a collision varies inversely with the mass of the particle. Since an electron has about 2,000 times less mass than a proton (and 8,000 times less mass than a helium nucleus), the electrons in the "plasma soup" are far more effective in exchanging energy with incident photons.

2. All matter produces photons of "blackbody radiation". Objects at room temperature produce long-wavelength infrared radiation, which is invisible to our eyes (but can be seen with night vision goggles). When objects are heated to a thousand degrees, they begin to glow red. A plasma at about 6,000 degrees (like the surface of the sun) glows white, emitting visible light, infrared and ultraviolet radiation.

   In the plasma of electrons, protons and helium nuclei at the surface of the sun, blackbody radiation is emitted by fast-moving electrons as their path is deflected by positively charged nuclei. Such "bremsstrahlung radiation" is produced whenever charged particles are accelerated.

10. Power from nuclear fusion

The fusion of hydrogen nuclei (protons) into helium nuclei provides the energy source for our sun and all other "main sequence" stars. This is one of the most energetic processes that occurs in the universe, so the hydrogen on Earth (mainly contained in water) could – in principle – provide all of mankind's energy needs for millennia. Consequently, for more than 75 years, scientists have been trying to "tame" the fusion of hydrogen so that it could produce heat and electricity used by mankind.

Fusion of "normal hydrogen" is relatively slow, even at temperatures of about 15 million degrees within the core of the sun. This is no problem for the sun, in which hydrogen nuclei might smash into each other again and again and again over billions of years before they finally get around to react, but power generation on Earth needs to occur over a much shorter timescale if it is to be practical. We humans are very impatient creatures, and investors want to get a return on investment in well under a billion years.

Consequently, research into fusion power generation has focussed on the reaction between the two other isotopes of hydrogen – deuterium and tritium. The reaction produces a helium nucleus and a neutron.

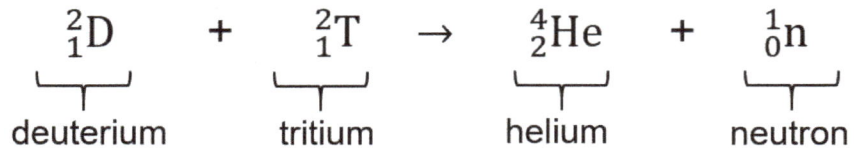

$$\underbrace{{}^{2}_{1}\text{D}}_{\text{deuterium}} + \underbrace{{}^{2}_{1}\text{T}}_{\text{tritium}} \rightarrow \underbrace{{}^{4}_{2}\text{He}}_{\text{helium}} + \underbrace{{}^{1}_{0}\text{n}}_{\text{neutron}}$$

About one in eight thousand hydrogen atoms is deuterium, so there is more than enough deuterium in the oceans to produce power by nuclear fusion. But it's a different story for tritium. Tritium has too many neutrons to be a stable nucleus. It undergoes radioactive decay (by emitting an electron) with a half-life of about twelve years, so tritium does not occur naturally on Earth. However, tritium can be produced artificially, and I will shortly discuss how we can use the fusion reaction itself to produce tritium.

The deuterium-tritium reaction is very unusual in one respect. Many nuclear reactions produce neutrons, but the deuterium-tritium reaction produces neutrons with extremely high energy. In fact, most of the energy liberated by the deuterium-tritium fusion reaction is imparted as kinetic energy to the neutron.

Energy liberated by the reaction pushes the helium nucleus and the neutron away from each other. It is exactly analogous to firing a bullet from a gun. When a gun is fired, energy released by burning gunpowder drives the bullet down the barrel – and drives the gun backwards. The conservation of momentum requires that the forward momentum of the bullet is equal to the backwards momentum of the gun. Energy is partitioned between the bullet and the gun in inverse proportion to their relative masses. Typically, the mass of a gun is several hundred times as much as the mass of the bullet it fires. This is why soldiers are not killed by the recoil of their rifle when they shoot at an enemy. This is why battleships had huge 16-inch cannons that could fire a shell weighing up to a ton, but destroyer escort vessels only had cannons that fired

Helium nucleus

Bam!!!

neutron

shells of much smaller calibre. It would be possible to build a small ship with a huge 16-inch cannon, but it would only get to fire one shot!

A helium nucleus has a mass of four atomic mass units (amu), which is four times the mass of a neutron. Consequently, the neutron produced by the deuterium-tritium fusion reaction gains four times as much kinetic energy in the "forward" direction as the helium nucleus gains in the "backwards" direction. In other words, four-fifths (80%) of the energy released by the fusion reaction is imparted to the neutron.

Like the fusion of "normal" hydrogen in the sun, fusion of deuterium and tritium requires incredibly high temperatures for the nuclei to approach each other with sufficient speed to overcome repulsion of their positive charges. To produce energy at a practical rate in a fusion reactor requires temperatures in excess of one hundred million degrees! The containment of such a hot plasma poses enormous technical challenges. If the plasma touched the wall of the container, the plasma would immediate cool and the reaction would be "quenched".

The major strategy to contain to contain extremely hot plasma in experimental fusion reactors is to use magnetic fields. Since deuterium and tritium nuclei (and electrons which surround them in the plasma "soup") have an electrical charge, their path is deflected by a magnetic field. Because the particles are moving so fast, the magnetic field must be extremely intense to maintain the nuclei within circular paths. Most experimental reactors contain the plasma within a "toroidal" (donut-shaped) vessel, from which air has been pumped out. Outside the toroidal vessel are located superconducting magnets. This basic design is based on early Russian research in the 1950s, and is referred to a "Tokomak".

Trapping a plasma within a magnetic field is actually a lot more complicated than it sounds. Don't forget that moving charged particles generate their own magnetic field, so the plasma tends to be unstable and eventually breaks out of the magnetic trap.

Also, even if it is confined, the plasma – like anything heated to extremely high temperatures – radiates "blackbody radiation". A metal bar heated to, say, a thousand degrees will glow red-hot. The surface of the sun, at 6,000 degrees K, glows white hot. A plasma at 100 million degrees glows with x-rays. The higher the temperature, the more intense is the emitted radiation, so super-hot plasmas will radiate energy intensely and cool very quickly. Unless energy loss by radiation can be replaced by intense heating, it will not be possible to heat the plasma hot enough for fusion to begin. Energy loss by emission of x-rays occurs mostly at the outside surface of the toroidal-shaped plasma. X-rays emitted *within* the plasma are mostly absorbed, with their energy retained within the plasma. So, to minimise radiative heat loss, the plasma must have a small surface area relative to its volume, and this is achieved by having a very large toroidal-shaped volume of plasma. In other words, for a Tokomak fusion reactor to produce more energy than is required to keep the plasma hot, it must be *BIG*.

During the early years of fusion research, in the 1950s and 1960s, experimental fusion reactors fit on a laboratory benchtop, but the energy required to heat the plasma (and keep it hot) was thousands of times more than the energy produced by the fusion reactor. As the technology developed, experimental fusion reactors got bigger and bigger. They also became more and more expensive, so research was increasingly undertaken jointly, with a number of countries contributing funding. Around 1981, when I was living in England, I got to visit the Joint European Torus (JET), an experimental Tokomak fusion reactor operated jointly by several European countries. At the time, JET was the most advanced experimental fusion reactor in the world. It was located near Oxford. I remember that there was some controversy at the time because English scientists were working side-by-side with German scientists, whose salaries were three times as much!

Since that time, research has progressed to the next iteration, which is now under construction in France. This too is a multi-country collaborative venture, called ITER. If successful, ITER will be the first fusion reactor that produces a net output of energy (that is, where energy released by the fusion reactor is more than the energy required to heat the plasma and operate the reactor). It is designed to produce about 500 megawatts of heat (produced when high-energy neutrons are absorbed by material surrounding the reactor). ITER will require 50 megawatts of electrical power to operate its reactor.

ITER is an experimental reactor, intended to show that a commercial fusion reactor is a practical next step. There is currently no plan to convert its heat output to electricity, as there is no need to prove that this would work (this technology is used in all coal-fired power stations). In typical coal or conventional (fission) nuclear power stations, heat output is converted into electricity with an efficiency of 30-35%. This indicates that the 500 megawatts of heat to be produced by ITER could generate about 150 megawatts of electricity. Subtracting the 50 megawatts required to operate the ITER reactor, this leaves a potential net electricity production of about 100 megawatts. That is less, but within the same scale, as the 350 megawatt units in conventional coal and nuclear power plants.

An excellent 5-minute video explaining ITER project can be viewed at:
https://www.youtube.com/watch?v=cCkp2SEsfao
This video was produced in 2014, so construction would have progressed since then.

ITER is an international collaborative project involving 35 countries, including most technologically-advanced nations. Although Australia was not one of the original member countries, Australia joined the project in October 2016, which will involve plasma researchers from the Australian Nuclear Science & Technology Organisation (ANSTO), Australian National University, University of Sydney, Curtin University, University of Newcastle, University of Wollongong and Macquarie University.

The ITER project is reported (https://en.wikipedia.org/wiki/ITER) to have already cost $14 billion by 2015. By the time it is finished, it will probably cost about 100 times as much as a coal-fired power station of the same electrical generating capacity. However, ITER is an experimental prototype. It is hoped that ITER will provide the experience and design data to enable the next generation of fusion power stations to be produced at much lower cost. Bear in mind that about $400 billion is invested each year in new electrical generating capacity around the world. In that context, a one-off $20 billion bet on a new technology that could radically change power generation doesn't seem unreasonable.

For ITER to successfully lead to succeeding generations of nuclear fusion power stations, it will be essential to obtain sufficient amounts of tritium fuel. The plan is to produce the tritium "in-situ" at nuclear fusion power stations, using high-energy ("fast") neutrons produced by the fusion reaction to irradiate lithium. The main reaction is:

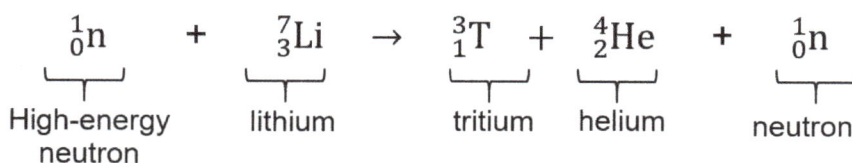

$$\underset{\text{High-energy neutron}}{\underbrace{{}^{1}_{0}\text{n}}} \quad + \quad \underset{\text{lithium}}{\underbrace{{}^{7}_{3}\text{Li}}} \quad \rightarrow \quad \underset{\text{tritium}}{\underbrace{{}^{3}_{1}\text{T}}} \quad + \quad \underset{\text{helium}}{\underbrace{{}^{4}_{2}\text{He}}} \quad + \quad \underset{\text{neutron}}{\underbrace{{}^{1}_{0}\text{n}}}$$

This reaction requires an energy input. The nuclear binding energy holding the tritium and helium nuclei together is less than the nuclear binding energy of the lithium nucleus. The reaction is driven by the extremely high energy of the impacting neutron. The reaction produces another neutron to replace the one that reacts, but the neutron produced has much less energy.

Some researchers, entrepreneurs and investors think that ITER is likely to be too big, complex and costly to lead to commercially-viable fusion power generation. A number of private companies and government-funded laboratories are developing alternative technologies for heating and confining hydrogen plasmas to achieve fusion on a smaller scale [Reference 1]. They are investing tens or hundreds of millions of dollars (rather than billions) to build and test prototypes to prove these concepts. These investors know that these technologies are "long shots" that are likely to fail, but could have a huge pay-off if they succeed. Their strategy is to either prove within a few years that they can achieve nuclear fusion, or to fail before they have invested huge sums. These may be the technologies that radically change our world in the 21st century, or they may disappear into the mists of history.

The Tokomak approach is not the only technology being developed with the ultimate aim of achieving a commercially-viable nuclear fusion reactor. Another strategy, undertaken by the US and French governments, is called "Inertial Confinement Fusion", (ICF). In this approach, huge lasers (the largest in the world) produce unimaginably intense pulses of light which are focussed onto a tiny pellet (less than one millimetre diameter) containing deuterium and tritium. The intense blast of light creates a shock wave that compresses the pellet to a tiny fraction of its original size and heats it to more than 100 million degrees. The fusion reaction produces a blast of energy, and the result is effectively a super-miniature thermonuclear bomb. By blasting perhaps one pellet every second, the plan is to achieve a near-continuous production of heat energy.

Inertial confinement research has been underway for decades at the US "National Ignition Facility". This facility seems to be strongly involved in military research on thermonuclear weapons (which use the deuterium-tritium fusion reaction to enhance the explosive yield of nuclear bombs), and this would likely be its main function. The technology for Inertial Confinement Fusion has been progressively scaled up, with each succeeding generation using larger and more powerful lasers. It is suggested (https://en.wikipedia.org/wiki/Inertial_confinement_fusion) that this technology has reached the milestone where the energy produced by fusion of a deuterium-tritium pellet was greater than the energy needed to fire the lasers. Another inertial confinement fusion (ICF) reactor of similar size has recently been built in France. This facility is probably linked to France's determination to maintain its own independent nuclear weapons capability. Inertial Confinement Fusion is one example where the same technologies and capabilities could be used for nuclear power generation and for nuclear weapons, and where a civilian nuclear power program could be used as a "cover" for nuclear weapon development and production.

A somewhat dated (1985) video about Inertial Confinement Fusion can be viewed at:
https://www.youtube.com/watch?v=vBKFoReQjyo

Reference
1. Alternative technologies being developed for nuclear fusion are discussed in the article "The fusion underground", Scientific American, Vol 315, No. 5, p. 32-39, November 2016.

11. Dawn of the nuclear era

For centuries, alchemists tried to produce gold from other elements. They mixed all sorts of exotic materials together, heated their mixtures and potions, stirred, and used other chemical methods (as well as making incantations to the spiritual world), but completely failed in their objective.

However, by the 1930's, physicists had discovered that they could "transmute" one element into another by irradiating its atoms with high-energy neutrons. Each neutron became bound to a nucleus, which then often emitted a Beta particle (a fast electron). The net effect was to add a proton to the nucleus (increase its atomic number by one). In effect, they were simulating the process that occurs in the final moments of giant stars, as they blow apart in supernova explosions, which produce an incredibly intense burst of neutrons. Neutron irradiation during supernova explosions is believed to have produced all elements more massive than iron (with 26 protons in their nucleus) – which includes many elements that are naturally found on Earth, including gold, silver and uranium.

In 1934, the famous physicist Enrico Fermi was investigating the transmutation of uranium into plutonium. As it turns out, this reaction is very important.

Uranium consists primarily of the isotope uranium-238. That's now. When the Earth first formed about 4.5 billion years ago from a disk of dust and gas, uranium atoms consisted of about equal amounts of the isotopes uranium-238 and uranium-235. However, the half life of U-235 is 700 million years, so in the 4.5 billion years since the Earth formed, the concentration of U-235 has reduced to 0.7%.

When uranium-238 is bombarded with neutrons, it initially forms the isotope Uranium-239. This isotope is very unstable, and ejects a beta particle (fast electron) to form a nucleus of Neptunium (named after the planet Neptune). This nucleus is also very unstable, and ejects another electron to form Plutonium-239. Here is the three-step nuclear reaction

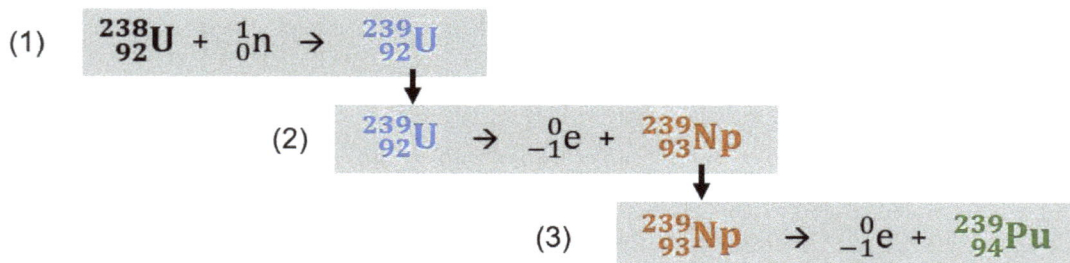

(1) $\quad ^{238}_{92}\text{U} + ^{1}_{0}\text{n} \rightarrow {^{239}_{92}\text{U}}$

(2) $\quad ^{239}_{92}\text{U} \rightarrow {^{0}_{-1}\text{e}} + {^{239}_{93}\text{Np}}$

(3) $\quad ^{239}_{93}\text{Np} \rightarrow {^{0}_{-1}\text{e}} + {^{239}_{94}\text{Pu}}$

However, when Fermi irradiated the isotope Uranium-235 with neutrons, he observed an entirely different process! The neutron did not become bound to the uranium nucleus but, instead, caused the uranium nucleus to split into two (or, sometimes, three) smaller nuclei. These smaller nuclei are much more strongly bound by the nuclear force than is uranium, so the process liberates an enormous amount of energy (about 200 million electron volts). This process was later named "nuclear fission".

What drives nuclear fission is that the large uranium nucleus is relatively unstable, and has far less binding energy than the "daughter nuclei" that are formed. The process is not very specific. A uranium nucleus is so keen to split apart that it forms just about any combination of mid-size nuclei (just so long as both "daughter nuclei" together contain 92 protons). Often, the isotopes that are formed are unstable and radioactive. Splitting of the uranium nucleus also produces 2 or 3 neutrons. Typically, for example, the fission reaction might produce barium and krypton radioisotopes, as follows:

$$^{235}_{92}U + ^{1}_{0}n \rightarrow ^{141}_{56}Ba + ^{92}_{36}Kr + 3\,^{1}_{0}n$$

Fission reactions have two characteristics of fundamental significance:

1. The reaction produces a huge amount of energy, which is mostly imparted as kinetic energy of the "daughter nuclei" as they fly apart, although a lot of energy is also imparted to the neutrons produced by the reaction.

2. While the reaction is "triggered" by a neutron, it produces more neutrons than it consumes. This provides a potential means to control the reaction. It also provides a means to have an uncontrolled chain reaction, where the reaction occurs at an exponentially increasing rate.

Initially, what motivated development of nuclear fission was the perceived need to develop an atomic bomb during World War II. Many leading physicists of the early 20th century were German, and the US and its allies were extremely concerned that Nazi Germany would be the first to develop an atomic bomb. If so, this could have had catastrophic consequences and changed the outcome of World War II. However, many leading German physicists had fled Germany in the 1930s and, as was later discovered (after Germany had been defeated), German efforts to develop an atomic bomb made little progress. Scientists leading the German atomic bomb project had no clear idea of how to achieve their objective. Instead, the Nazi regime devoted major resources to developing rockets and jet engines. By contrast, the US and its allies devoted massive resources to the "Manhattan project" to develop an atomic bomb.

Major challenges needed to be overcome to achieve near-instantaneous release of the energy produced by the fission reaction. To get a run-away chain reaction, each uranium-235 nucleus undergoing fission needs to produce sufficient neutrons to trigger at least one further fission reaction. This would appear to be simple since various fission reactions produce an average of about 2.5 neutrons. However, if the mass of uranium-235 is insufficient, many neutrons escape from its surface without striking another uranium-235 nucleus and triggering a fission reaction. Many atomic nuclei absorb neutrons (including, as we have seen, uranium-238), and this reduces the number of neutrons available to initiate further fission reactions.

For the fission reaction to be self-sustaining, the lump of uranium-235 must exceed a certain "critical mass". Below this threshold, in a "sub-critical mass", most of the neutrons produced by fission reactions escape or are absorbed. Remaining neutrons cause fewer and fewer fission reactions to be triggered. In this way, the number of fission reactions would undergo an exponential decay and would "fizzle out".

This process is very similar to the outbreak, or the containment, of epidemics of infectious diseases (like the flu, whooping cough, polio, AIDS and other viral infections). Once a person is infected, the virus replicates within the "host" and (for a period of time while the person is infectious) is transmitted to other victims by coughing, sneezing, food contamination, contact with contaminated surfaces or sex. If each infected person passes the disease to at least one other person, the disease will spread through the population and spiral exponentially into an epidemic. If each infected individual transmits the disease to less than one other person, the epidemic will reduce and "fizzled out". This is why vaccination can be so effective in preventing or halting epidemics. Even if a vaccine is not 100% effective in preventing infection, and even if 100% of the population is not vaccinated, an epidemic will be avoided or "fizzle out" if, on average, less than one new infection occurs for each person that is already infected.

In the same way, the fission reaction can only spiral into an explosive chain reaction if the uranium-235 (or other "fissile" radioisotope) exceeds a critical mass. The critical mass depends on the concentration of U-235. Uranium within the Earth's crust contains only 0.7% U-235, and so, would not sustain an uncontrolled chain reaction. The other isotope, Uranium-238, which comprises 99.3% of uranium nuclei, does not readily undergo fission, and even absorbs neutrons. Consequently, to make an atomic bomb using uranium, Uranium-235 must be concentrated (or "enriched") a hundred-fold.

Producing "highly enriched" U-235 is technically very difficult – which is fortunate for humanity – as this has been the major barrier preventing other countries from making their own nuclear bombs. U-235 has the same chemical properties as U-238, so the two isotopes cannot be separated by chemical means. The mass of each U-235 atom is about 1% less than U-238 atoms, and this difference in mass is utilised to separate the two isotopes (through many stages of centrifuging).

Once highly-enriched U-235 was obtained, it was relatively straightforward to assemble the atomic bomb that was dropped on the Japanese city of Hiroshima. This bomb was called "Little boy" because it was long and thin (for reasons that will soon become apparent). It was only necessary to bring together two pieces of U-235 with sub-critical mass, to form a larger piece that exceeded the critical mass. However, the critical mass of U-235 had to be formed extremely quickly. Otherwise, as soon as the fission reaction started, heat released by the reaction would vaporise and blow apart the uranium – stopping the fission reaction.

The "Little boy" bomb was reportedly made from an artillery cannon barrel, which fired a projectile containing a sub-critical mass of U-235 into a second sub-critical mass of U-235. Shooting the two pieces of U-235 together in this way enabled about 1.5% of the U-235 to undergo fission before the material was blown apart. Even though the explosion was inefficient and small by the current standards of nuclear weapons, the fission reaction produced temperatures exceeding 100 million degrees (ten thousand times hotter than the surface of the sun). The near-instantaneous release of heat produced an explosive blast of the same destructive power as 15,000 tonnes of conventional TNT explosive.

The bomb, which exploded at an elevation that would cause maximum damage, vaporised material within the underlying blast zone. This material was irradiated by the burst of neutrons produced by the fission reaction, forming radioactive isotopes that were carried to the upper atmosphere. The vaporised material (combined with radioisotopes formed directly by the fission reaction) condensed into fine dust particles, forming a characteristic "mushroom cloud", which drifted back to the Earth's surface in the following hours and days.

"Little boy" bomb being loaded into a B-29 bomber. Source: US National Archives.

By July 1945, the United States had produced only enough enriched U-235 to produce one atomic bomb. To make sufficient "fissile" material to produce numerous atomic bombs, the Manhattan project also developed the capability to sustain a controlled nuclear reaction (a nuclear reactor). The fission reaction inside the reactor produced heat energy at a controlled rate, which could be used to generate electric power (as now occurs in hundreds of nuclear power plants around the world). The fission reaction also produces neutrons, which are absorbed by uranium-238 to produce plutonium-239 (via the transmutation reaction discussed earlier). Plutonium-239 is also "fissile". Some Pu-239 produced within a reactor undergoes fission, contributing to the energy (and neutrons) produced by the reactor. But some Pu-239 is retained within the "spent fuel" produced by the reactor, from which it can be extracted.

The first atomic bomb that was tested in the desert of Nevada in 1945, as well as the bomb dropped on Nagasaki, employed plutonium-239. These bombs used a different mechanism than the uranium-based "Little Boy" bomb to achieve a critical mass. A hollow sphere of plutonium-239 (the "bomb core") was surrounded by conventional high explosives. By detonating these explosives in a precisely controlled, symmetrical pattern, the hollow shell of plutonium was imploded into a compact spherical ball. The compact shape of the plutonium caused the critical mass to be exceeded. Because of its bulbous shape, this bomb was called "Fat man". Three days after the

The "Fat man" plutonium bomb which destroyed Nagasaki: Source: US National Archives

city of Hiroshima was destroyed by "Little boy", the "Fat man" bomb was dropped on the city of Nagasaki, releasing the explosive power of 21,000 tonnes of TNT.

Within days of the destruction of Nagasaki, Japan surrendered, ending the second World War. At that time, the United States had the capacity to build 2-3 atomic bombs per month. The development, manufacture and deployment of nuclear weapons intensified in the post-war years to meet the threat posed by the Soviet Union, which had developed their own nuclear weapons within a few years (aided by spies working within the Manhattan project). By the early 1960s, the US and Soviet Union had produced and deployed tens of thousands of nuclear bombs. Some were carried on B-47 and B-52 bombers (and Soviet equivalents) maintained in constant in-flight readiness, within missile silos and mobile missile launchers, and as warheads on submarine-based missiles. The US, Soviet Union and other nuclear-armed countries developed more powerful, smaller and more mobile nuclear weapons – including "thermonuclear weapons" that utilise fusion reactions to enhance the yield of fission bombs. Many of these bombs have fifty times more explosive power than the bombs which destroyed Hiroshima and Nagasaki.

Both the US and Soviet Union developed "tactical nuclear weapons" (smaller bombs intended to destroy formations of enemy troops or tanks, fleets of ships, nuclear missiles or military installations) and "strategic nuclear weapons" (larger bombs intended to destroy cities). Through the 1950s, the intention of military planners was to use tactical nuclear weapons to win the next war. The role of "strategic nuclear weapons" was to deter the enemy (Soviet Union) from starting a nuclear war, since launching a nuclear attack would lead to retaliation that would devastate the attacking country (and perhaps all humanity). The major nuclear powers soon realised that the use of "tactical nuclear weapons" was not a viable strategy to "win wars". An exchange of even "small" nuclear weapons would rapidly escalate into a global apocalypse, and a major nuclear war would have no "winners". The strategy of "Mutual Assured Destruction" (MAD) was used to mediate the military stand-off between the United States and Soviet Union during the Cold War. As one commentator famously noted "If World War 3 is fought with nuclear weapons, then World War 4 will be fought with stones and spears".

The break-up of the Soviet Union and easing of Cold War tensions reduced, but has not eliminated, the threat of nuclear annihilation. Thousands of nuclear weapons are still deployed by military forces of the US, Russia, China, Britain, France, India, Pakistan, probably Israel and likely soon by North Korea. A change in the global strategic outlook could spur other countries to develop their own "nuclear deterrence". The more countries have nuclear weapons, the greater the scope for them to be used deliberately or by miscalculation in regional power struggles, and the greater the chance that uranium or plutonium might be stolen by terrorist groups. Pakistan, which has fought several conventional wars against India, is reported to be developing tactical nuclear weapons. The violent and unstable situation in the Middle East, including the bloody civil war and use of chemical weapons in Syria, shows that ruthless and fanatical leaders would be prepared to use weapons of mass destruction.

Next, we'll be talking about how nuclear fission can be controlled, providing a major source of electrical energy with minimal greenhouse gas emissions. The civilian nuclear power industry, which provides a significant share of the world's electricity with minimal greenhouse gas emissions, originated from military research to develop nuclear weapons. Nuclear power generation and nuclear weapons are two sides of the same coin. They are inextricably linked. They share the same technologies, materials and capabilities. Many countries which operate nuclear power reactors do not have nuclear weapons, but all countries that have (or are developing) nuclear weapons operate nuclear reactors.

Many attempts have been made over past decades to reduce the number of nuclear warheads and limit their proliferation, but the elimination of nuclear weapons may be unachievable. While military powers might hopefully be prepared to reduce the number of their nuclear weapons or limit their deployment, it seems inconceivable that any nuclear-armed country would eliminate their arsenal while rival powers might be holding a concealed arsenal. The only viable strategy for the US and its allies to deter North Korea from destroying Soul, once it has developed a nuclear weapon that can be carried by a missile, is by retaining the capability to instantly retaliate and annihilate that country and its leadership. Once technology has been developed to produce nuclear weapons, it cannot readily be "unlearned". It may be impossible to "put the nuclear genie back into the bottle".

A good overview of the development of nuclear weapons is provided by the following four-minute video on "How nuclear weapons work":
https://www.youtube.com/watch?v=E-Ks-Bs0MPM   ("0" = zero)

A sobering account of the apocalyptic effects of a nuclear weapon on a major city like New York is given in the following brief (four minute) video:
https://www.youtube.com/watch?v=Aza-2wopCFY

12. Nuclear power reactors

To generate electricity in a nuclear power reactor, we need to control the fission reaction so that it neither fizzles out, nor runs away. The fission reaction needs to operate at the point where it is just "critical", that is, where neutrons produced by a nucleus of uranium-235 undergoing fission cause *one* other U-235 nucleus to undergo fission – no more, no less.

The first challenge is to get the reaction to sustain itself and not fizzle out. High-energy neutrons produced by the fission reaction (with an energy of about one million electron volts) are not very efficient in causing fission to occur. In an atomic bomb, this problem is overcome by "brute force" - using very highly enriched U-235 (or pure plutonium-239). Then, even though high-energy neutrons are weakly absorbed, the high concentration of U-235 ensures that enough neutrons are absorbed to get a run-away fission chain reaction. However, for a practical nuclear power reactor, we want to be able to use un-enriched uranium (0.7% U-235) or slightly enriched (2-3%) uranium-235.

It turns out that slow neutrons are much more strongly absorbed by U-235 nuclei, and are much more effective in causing the fission reaction. To get strong absorption, we need to reduce the kinetic energy of the neutrons by at least ten million times, and reduce their velocity by at least 3,000 times. This is done by bouncing the neutrons off other nuclei. During each collision, a neutron transfers some of its kinetic energy to the other nucleus. After thousands of such collisions, the neutron is in "thermal equilibrium" with the other nuclei, having the same average kinetic energy as any atom at room temperature (about 0.04 electron volts).

The ideal material to use as a "moderator" to slow the neutrons should have atomic nuclei with relatively low mass. If the neutron and moderator nuclei have similar mass, a high proportion of the neutron's energy is transferred during each collision. It is also very important that the moderator material does not absorb neutrons, since each neutron undergoes thousands of collisions with moderator nuclei before it strikes another nucleus of U-235. The ideal moderator should also be relatively cheap, non-toxic and available in pure form (since many impurities, even at low concentrations, will strongly absorb neutrons).

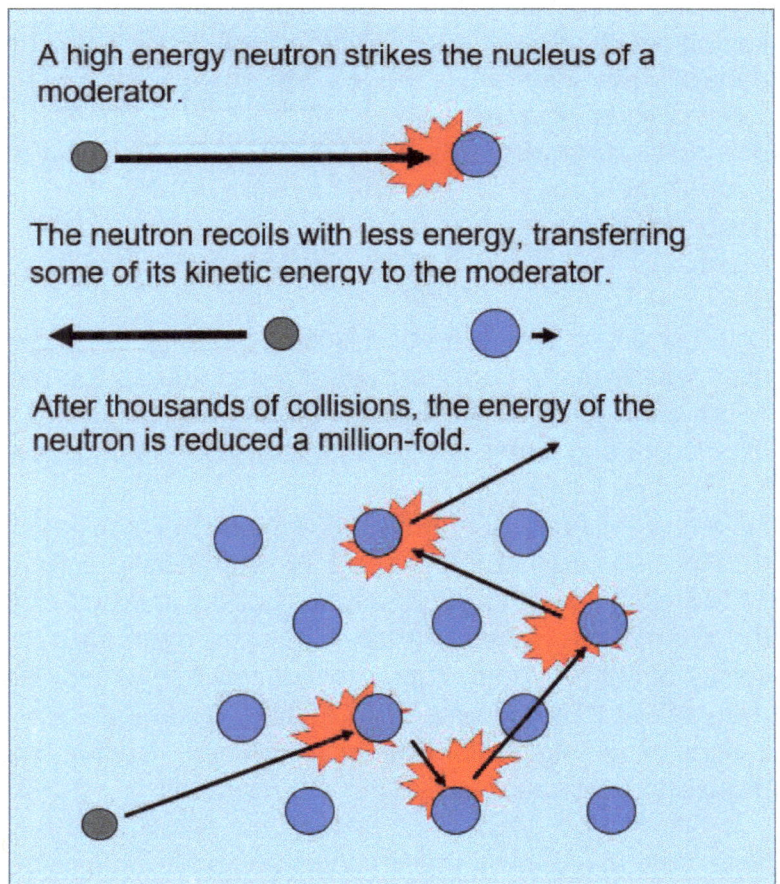

A high energy neutron strikes the nucleus of a moderator.

The neutron recoils with less energy, transferring some of its kinetic energy to the moderator.

After thousands of collisions, the energy of the neutron is reduced a million-fold.

Moderating the energy of the neutrons is the "key" to a controlled fission reaction in a nuclear power reactor, avoiding the need for highly-enriched uranium. However, this is not the case for the uncontrolled fission reaction in a nuclear bomb. It literally takes some time (perhaps a thousandth of a second) for a neutron to bounce off thousands of moderator nuclei, losing velocity in each collision. This is no problem in a nuclear power reactor, which is operating with a continuous power output. However, in an atomic bomb, the fission chain reaction must occur within about **a billionth of a second** before the material blows apart.

One common moderator is normal hydrogen, contained in water. Hydrogen is ideal in that its nucleus (a single proton) has very low mass, although hydrogen does have a slight tendency to absorb neutrons, and this is compensated by using slightly enriched uranium. Water is used as a moderator in several types of reactor, including the Boiling Water Reactor (BWR). In BWRs, fuel rods (uranium pellets contained within zirconium tubes) are surrounded by liquid water, which boils into high-pressure steam within the reactor. The steam is used directly to drive a spinning turbine which, in turn, spins an electrical generator to convert mechanical power to electricity (exactly as is done in a coal-fired power station). After the steam expands in the turbine, it condenses back into water and is pumped back into the reactor. The rate of the nuclear fission reaction is controlled by inserting "control rods", made of a material which strongly absorbs neutrons, into the reactor.

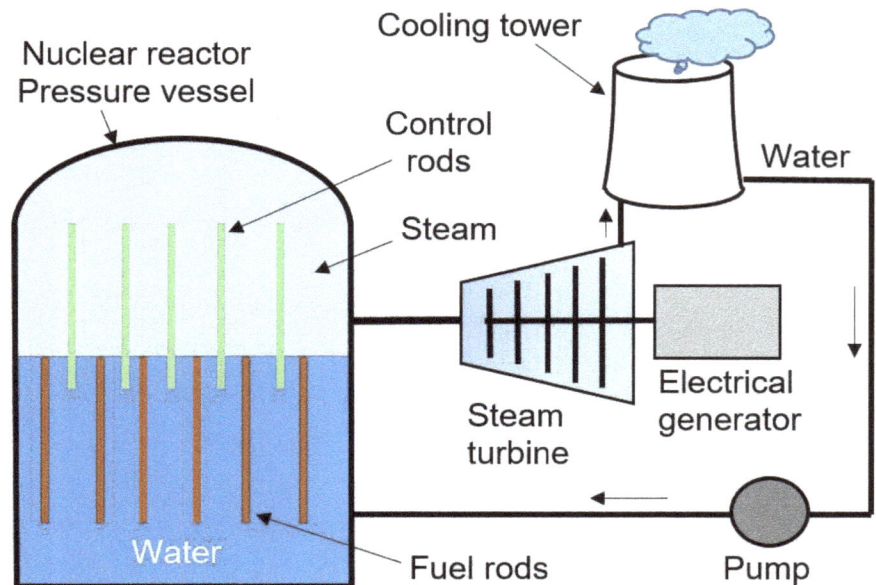

Schematic diagram, showing the main components of a Boiling Water Reactor power station.

During normal operation of a Boiling Water Reactor, water is pumped into the reactor at the same rate as it boils into steam and leaves the reactor. The water serves as a coolant, "working fluid" and moderator. This provides a safety feature in that, if there is a catastrophic loss of cooling water, the reactor loses its moderator and this slows or stops the reaction.

However, this safety feature did not avert catastrophic failure in the boiling water nuclear reactor near Fukushima, Japan. In response to a major earthquake off the coast of Japan on 11 March 2011, automatic safety systems lowered control rods to stop the fission reaction at the Fukushima Power Station. However, even after the fission reaction ceases, radioactive decay of short-lived isotopes (produced by the fission reaction) continues to release heat (equivalent to about one-tenth of heat production during full power output). Failure of the pumps providing cooling water caused the reactor core to overheat, perhaps reaching several thousand degrees and melting the fuel rods.

Scientists discovered that a naturally-occurring Boiling Water Reactor occurred about 1.7 billion years ago within a uranium ore deposit in Gabon, Africa. At that time, U-235 comprised about 3% of uranium isotopes (about the same concentration currently used in most nuclear power reactors). Ground water permeated into the porous uranium deposit, acting as a moderator, causing the fission reaction to become critical. Heat released by the reaction caused the water to flash into steam and escape. Loss of the moderator caused the fission reaction to cease. After several hours, the deposit cooled and, once again, water seeped back into the porous rock. In this way, the fission reaction pulsed on and off every three hours or so, for perhaps thousands of years.

Another common reactor design is the Pressurised Water Reactor (PWR). The PWR was originally developed for application in nuclear-powered submarines and aircraft carriers by the US Navy, but has since been widely adopted for commercial nuclear power stations. The pressurised water reactor also uses water as a moderator and coolant, although water circulating through tubes in the reactor is prevented from boiling by high pressure. The water in this "primary circuit" is passed through a heat exchanger, where water in a "secondary circuit" boils into steam. This steam is then expanded in a turbine to produce power, as in any other power station. The use of a heat exchanger ensures that water passing through the reactor (which may contain radioisotopes) is kept separate from the water/steam passing through the turbine. The PWR is also self-regulating. If the temperature inside the reactor rises, water inside the reactor becomes less dense and less effective as a moderator, causing the fission reaction to slow.

A good 5 minute video, showing how Pressurised Water Reactors work can be viewed at: https://www.youtube.com/watch?v=_UwexvaCMWA

Boiling Water Reactors and Pressurised Water Reactors are the main types of nuclear power reactors in operation around the world, although other variations have been developed. RBMK reactors operating in Russia and former Soviet republics use a graphite moderator. This includes the reactor which exploded near Chernobyl in the Ukraine in 1986. The dispersal of radioactive debris during the run-away nuclear reaction and explosion was exacerbated by burning of the graphite moderator and by the lack of a reinforced concrete outer shell. Great Britain has developed an Advanced Gas Cooled Reactor, which uses a graphite moderator and pressurised carbon dioxide as a coolant.

The Canadians have developed their own nuclear reactor technology which is quite distinct from other common reactor types. It utilises deuterium, contained in heavy water, as a moderator. Deuterium, with its low mass and virtually no absorption of neutrons, is arguably the best moderator. The excellent moderator qualities of heavy water allow the Canadian reactor technology, called CANDU (**CAN**adian **D**euterium **U**ranium), to operate with un-enriched uranium.

A good 11-minute video about the Darlington CANDU nuclear power station can be viewed at: https://www.youtube.com/watch?v=_AdA5d_8Hm0

It is useful to consider the operational role of nuclear power reactors in an integrated electricity grid. Most countries now have extensive electricity grids, extending over large areas with millions of residents. integrating many power stations into a distribution system providing power to industrial, commercial and household consumers. The grid must match the operating characteristics of different types of electricity suppliers to meet the needs of consumers.

The electricity needs of consumers varies widely over a typical day, and from season to season. In most parts of Australia, electricity demand reaches a peak during hot summer weekdays, when businesses and households operate air-conditioning. During the night, and during holiday periods, most businesses are shut and electricity demand is much less.

The electricity grid must bring on additional generating capacity to meet demands during peak periods, while dispatching surplus generating capacity during off-peak periods. This requires juggling and horse-trading between suppliers and the grid operator, which is now done almost instantaneously through a computerised bidding process. Suppliers offering the cheapest bid are selected to supply power during each 30-minute period.

Roughly, here's how it works. Some generators, including solar and wind power stations, are "non-dispatchable". These produce electrical power whenever the sun shines or the wind blows, and they must accept whatever price is offered for power at that particular time. Since renewable power stations consume no fuel, their operating costs are the same whether they are producing power or not. Whatever revenue they earn by selling power to the grid is more than they would get by not selling power to the grid. The "non-dispatchable" aspect of renewable energy can be mitigated by dispersing the grid across a wide geographical area (so that, for example, wind generation in windy areas offsets lack of generation by wind farms beset by still conditions). It can also be reduced by installing energy storage technology, which absorbs and stores electrical energy when surplus power is available, and releases it during peak-demand periods. However, most types of energy storage technology are still expensive or are only practical in certain circumstances.

On the other end of the hierarchy are gas turbine power stations, which can be built relatively cheaply and can be started and shut-down quickly. However, gas turbine stations use relatively expensive fuel (natural gas or diesel) and are only profitable to operate when a premium price is offered through the electricity grid. Consequently, these stations are normally used to provide power only during peak-demand periods.

For economic and technical reasons, coal-fired power stations are operated as "base-load" power stations. They are relatively costly to build, but since their fuel (coal) is relatively cheap, their operating costs are low. Furthermore, starting-up and shutting-down large coal-fired boilers is time-consuming and problematic. Whenever the boilers are brought to operating temperature, the steam tubes expand (and later, contract, when the boiler is shut down) and are subjected to stress and (eventually) to cracking. Operators of coal-fired power stations want to keep their generating plants operating continuously (except for scheduled maintenance) – even when they must accept a low price for electricity produced.

So, where do nuclear power stations fit within this hierarchy? Nuclear power stations are "super base-load". They are the most expensive type of power station to build (partly because of high safety standards that are applied), and are the cheapest to operate. Nuclear plant operators want to keep their reactors at operating temperature all the time: the last thing they want is regular heat-up and cool-down cycles leading to cracks in boiler tubes.

Consequently, any debate on the "best type" of electrical generating technology needs to take account of how the requirements of electricity consumers varies over time, and how each type of generating technology can best be used to meet the varying load requirements.

13. The nature of reality, and our place in the universe

Introduction

In the 1950's post-war years, everything seemed possible. The introduction of antibiotics and new vaccines promised to eliminate disease and illness. Insecticides, herbicides and artificial fertilisers promised an unlimited bounty of food to maintain growing population and prosperity. The age of jet travel had dawned, making the world accessible to millions of travellers. Soon, it was projected, we would be travelling in flying cars or strapping rocket packs on our backs. Plastics and new metal alloys promised to overcome the limitations of materials. Nuclear energy was expected to produce unlimited amounts of power. For the first time, television allowed ordinary people to be entertained by the world's leading singers, dancers, actors and comedians, and we could see exotic places that had previously been restricted to our dreams. Launching of the first space satellites showed that man's domain need not be limited to the Earth, but could extend deep into space. Soon, men would walk on the moon, and many expected that astronauts would be living on space stations and other planets before long.

Some of these projections did come to pass, but by the early 1960s, many people realised that reality was not quite heading towards the space-age utopian vision of the future that had been portrayed.

The "Technology will solve all our problems" mindset began to look flawed. The idea that we could control and dominate nature began to show its dark side. It became apparent that nuclear weapons posed an existential threat to the survival of humanity, that vastly expanded industrial production was poisoning the environment, that many of the new materials would not degrade once they had been discarded into the environment, and that insect pests and bacteria were becoming resistant to the chemical agents developed to eliminate them. Pre-occupation with technological supremacy and control led to hubris which might have contributed to the United States and Australia becoming embroiled in a long, destructive and pointless war in Vietnam.

A counter-culture developed, and I think that its mood is best encapsulated by the protest song "The eve of destruction" in 1965:

> "You may leave here for four days in space, but when you return, it's the same old place".

Furthermore, the idea that we knew everything that we needed to know to control our destiny began to look flawed. Some fifty years before, Albert Einstein had showed in his Special and General Theory of Relativity that our basic assumptions about how the universe worked – things that everyone takes for granted as "common sense" – are fundamentally wrong.

The implications and consequences of Einstein's theories are truly astounding. I suspect that most scientists are still trying to come to terms with what it means. I know that I am. However, the rationale and derivation of Einstein's conclusions are relatively straightforward. Many people think that understanding the reasoning behind the theory of relativity is beyond the capacity of ordinary people, but I disagree. With little math beyond high school trigonometry, the basic arguments underlying Special Relativity can be explained and its conclusions derived. I set out to do exactly this on the following pages, and leave it to readers to judge whether I am successful.

Einstein's genius was not the ability to do complex calculations, but an ability to look at things from an entirely new perspective, applying logic and intuition to a series of imaginary "thought experiments". In the hundred years since Einstein's Special and General Theories of Relativity were published, its predictions have been thoroughly tested and shown correct in every case. Most recently, Einstein's prediction of gravity waves (the warping of space propagating at the speed of light, caused by massive bodies being accelerated) was confirmed. In 2016, two international teams detected gravity waves emitted 1.3 billion years ago by two black holes spiralling inwards before merging into a single black hole. "Gravity wave astronomy" is now expected to become a new technique to search for other such cataclysmic events in distant galaxies.

In the decades since Einstein proposed his theory of relativity, quantum mechanics was developed to explain the bizarre behaviour of atoms and sub-atomic particles. Quantum mechanics posits that an inherent degree of uncertainty constrains our ability to "know" the positions and velocities of these particles. In other words, below certain dimensions of size and mass, there are limits to what we (as an observer investigating the sub-atomic world) could *ever* know. Also, during the same period, astronomers discovered that they had under-estimated the scale of the universe – by a factor of billions!

Only in the 1950s and 1960s were the implications of relativity and quantum mechanics brought to the awareness of the general public. These developments provided a new outlook on the nature of reality, and our place in the universe. No longer were space and time separate entities - they are inextrinsically linked into four dimensions of space and time. The spacing, timing and sequence of events can be different for observers moving at different speeds. It was apparent that the universe is a much stranger place than we thought it was. Perhaps our scientific knowledge was driven by deeper laws of physics, of which we could now only guess. And while it was clear that our society had been extraordinarily clever at creating new technology, we often used this power in thoughtless, frivolous and selfish ways. Some people began to wonder whether technology would solve our problems if they stemmed from our own human shortcomings.

These unsettling questions were explored by the popular television drama series "The Twilight Zone". I remember that, at the beginning of each episode, Rod Serling (the originator and main writer of the series) would introduce the program. His introduction varied over the five years that the program ran, but started out like this:

> *"There is a fifth dimension beyond that which is known to man. It is a dimension as vast as space and as timeless as infinity. It is the middle ground between light and shadow, between science and superstition, and it lies between the pit of man's fears and the summit of his knowledge. This is the dimension of imagination. It is an area which we call the Twilight Zone."*

You can hear Rod Serling's voice, with the creepy background soundtrack, in an introduction to the "Twilight Zone" at: https://www.youtube.com/watch?v=NzlG28B-R8Y (l is lower case "L") It brings back my sense of wonder at what I don't understand, and what we don't know might be out there.

14. Special relativity, and the nature of reality

Throughout history, people have tended to view themselves as the centre of the universe. Until a few hundred years ago, most people thought that the sun, moon, planets and stars all revolved around the Earth. Galileo and other early scientists provided convincing evidence that the Earth, moon and planets revolve around the sun, but this did not go down well with the existing church establishment. Eventually, it became generally accepted that the Earth was one of eight or nine planets that orbit our sun, and that the sun has the central position in our solar system. However, by the early 1900s, astronomers realised that our sun was one of **billions** of stars within our galaxy. By the 1930s, it was becoming clear that our entire galaxy (now estimated to contain 400 billion stars) is merely one galaxy among billions. Our place in the universe was starting to look less and less unique and special.

It is natural for we humans to consider the Earth as a special and unique place, and as a reference point for everything that happens. It is the only planet on which we, our parents and all our ancestors have ever lived. We view everything that happens, or anything that moves, as being relative to the surface of the Earth or the air in its atmosphere. The Earth on which we live is our "frame of reference" for everything that happens in human experience.

In the latter half of the 19th century, a revolution was occurring in our understanding of light. Of huge importance was the work of James Clerk Maxwell, who showed that visible light is one type of electromagnetic radiation (as are radio waves, microwaves, infrared radiation, ultraviolet radiation, x-rays and gamma rays, whose vastly different properties are due to their enormous span of frequencies). Maxwell proposed that an alternating electric field (say, from a radio transmitter) induced a magnetic field which alternated at the same frequency, and that this changing magnetic field induced a changing electric field. In this way, energy is constantly interchanged between the electric and magnetic fields, causing light to propagate through space. Maxwell calculated that such an electromagnetic wave would move at a characteristic velocity (given the symbol c) which depended only upon the "electric permittivity" and "magnetic permeability" of free space. Both the permittivity and permeability of a vacuum had been measured experimentally to a high degree of accuracy, and when these numbers were inserted into Maxwell's equation, the calculated value for the speed of light was a perfect match with the experimentally-measured value (299,792,458 metres/second).

The fact that free space, or a vacuum, has a characteristic "electric permeability" and "magnetic permittivity" suggests that a vacuum is not simply a complete absence of everything. Scientists suggested that the entire universe is permeated by "aether", a medium that carries light – exactly analogous to the way that air acts a medium to carry sound waves.

Scientists who studied sound waves knew that sound moved through the air at a speed of about 300 metres/second. Water waves move at a certain speed relative to the ocean. Other types of waves moved through a medium, at a fixed speed relative to that medium. It was perfectly natural, even "obvious", to expect that light would move at a fixed velocity relative to the "aether" medium.

Since the "aether" pervades the entire universe, scientists reasoned, it should be possible to measure the speed that the Earth is moving through the aether. Two scientists, Albert Michelson and Edward Morley, set out to do exactly this. By 1887, they had constructed an ingenious and sophisticated apparatus and were ready to conduct experiments to determine the absolute speed of the Earth.

The Michaelson-Morley apparatus sent a beam of light to a half-silvered mirror, which split the light beam into two paths (shown in blue and red in the diagram). One light beam travelled distance **Sx** in one direction (which we'll call the x axis), reflected off a mirror, and returned the same distance **Sx** back to the half-silvered mirror. The other light beam travelled in the perpendicular direction (which we will call the y axis) for distance **Sy** (which was exactly equal to **Sx**), reflected off another mirror, and returned back to the half-silvered mirror. When the two light beams recombined, Michaelson and Morley reasoned, they would be slightly out-of-phase (depending upon the speed of the Earth through the aether). This instrumental technique is very sensitive (it's called an interferometer), and Michaelson and Morley determined that they should be able to measure a tiny change in phase of the two light beams – even using the relatively simple technology available at that time.

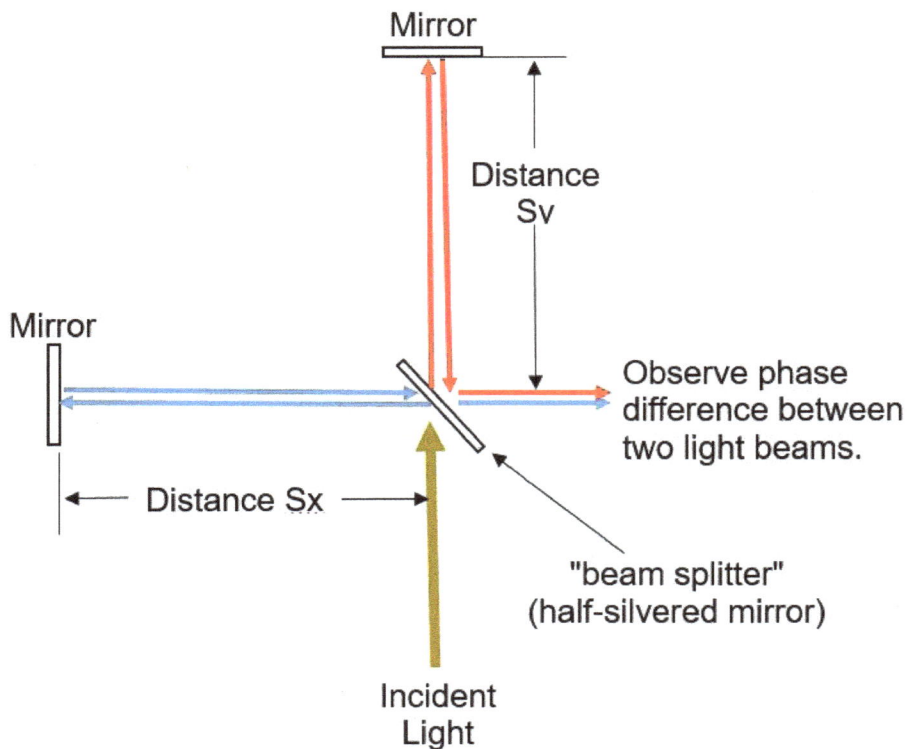

Mirror

Distance Sy

Mirror

Observe phase difference between two light beams.

Distance Sx

"beam splitter" (half-silvered mirror)

Incident Light

Michaelson and Morley reasoned as follows. If the Earth happened to be moving through the aether in the direction of the x-axis, then the light beam in the x-direction would move slower (as if it had a "headwind"). It would take more time than the other light beam to reach its mirror, but on the return journey, the light beam would be moving faster (as if with a "tailwind") and take less time than the light beam in the y-direction. The greater time for the outgoing journey, and the reduced time for the return journey, would not cancel each other. Overall, the light beam travelling in the x-direction would take longer to complete the journey, and would arrive later (and out of phase) with the light-beam moving in the y direction. The entire apparatus floated on a pool of mercury and could be rotated to align with the direction that the Earth was moving through the aether.

The Michaelson-Morley experiment was brilliant logic and engineering. Had the Earth been moving at even a modest velocity through the aether, their apparatus would have been sufficiently sensitive to detect it. Michaelson and Morley conducted the experiment many times, rotating the apparatus at all possible angles. But every time Michaelson and Morley conducted the experiment, they didn't detect any phase shift between the two light beams!

What could this mean? Does the Earth sit motionless in the aether? Is the Earth the centre of the universe after all? Does the Earth drag the aether with it as it moves through the universe?

This made no sense at all. By then, scientists were well aware that the Earth was rotating on its axis, with the Earth's surface moving west-to-east at 1,600 kilometres/hour at the equator, and changing direction every 12 hours. Furthermore, the Earth rotates around the sun, moving along a circular orbit at 30 kilometres/second, so the motion of the Earth through the aether should show a change in direction every six months. The Michaelson-Morley experiment was repeated at different times of the day and night, and throughout the year. The result was always the same. Both light beams arrived exactly in phase.

The Michelson-Morley experiment has been called "the most famous failed experiment in history". It was repeated by many researchers, using increasingly sophisticated and sensitive equipment. Every attempt failed to measure the velocity of the Earth through the aether.

Scientists were baffled by this result, and for years, no one could explain what it meant. The answer was provided by Albert Einstein in the Special Theory of Relativity in 1905. Einstein's genius was based on his ability to look at the universe in an entirely new way, and to not give up when he predicted results that were completely at odds with "common sense" (as it was then considered). His theory set out a new paradigm for reality.

Einstein reasoned that there is nothing special about the Earth's position or velocity in the cosmos. Indeed, the opposite is the case. If the Michaelson-Morley experiment were conducted on any other planet or in another galaxy, Einstein reasoned, exactly the same result would be obtained. Not only do we Earthlings not know our absolute velocity through space, we **cannot** know our absolute velocity – but nor can any other observer anywhere in the universe. We can only determine the **relative** velocity of various objects – how fast they are moving in relation to each other.

Einstein based his theory on two assumptions:

1. The same laws of physics apply everywhere in the universe.

2. The speed of light is the same for all observers, regardless of where they are in the universe.

Taking these assumptions to their logical conclusion led Einstein to develop a dramatically different understanding of the universe, led to unexpected and astounding predictions (which have since been confirmed), and turned some of our most deeply held beliefs on their head.

To follow Einstein's reasoning, let's say that we construct a spaceship containing the Michaelson-Morley experiment, and we launch it into space at some velocity **v** relative to Earth. The spaceship is manned by your cousin Fred, who you know to be completely honest and reliable. As Fred performs the experiment, he reports back on what he observes. Meanwhile, you carefully observe the experiment from planet Earth. A key difference in perspective arises because the light beam in the experiment will always move at velocity **c** **relative to your world on Earth**, and the same light beam will move at velocity **c relative to Fred's spaceship** (moving at velocity **v** relative to us).

First, let's consider the light beam moving in the same direction as Fred's spaceship (along the x-axis). According to Fred's observations, the time for the light beam to travel to the mirror and back is simply the round trip distance **2Sx** divided by the velocity of light **c**.

Fred's spaceship

Sy

Sx

Velocity v

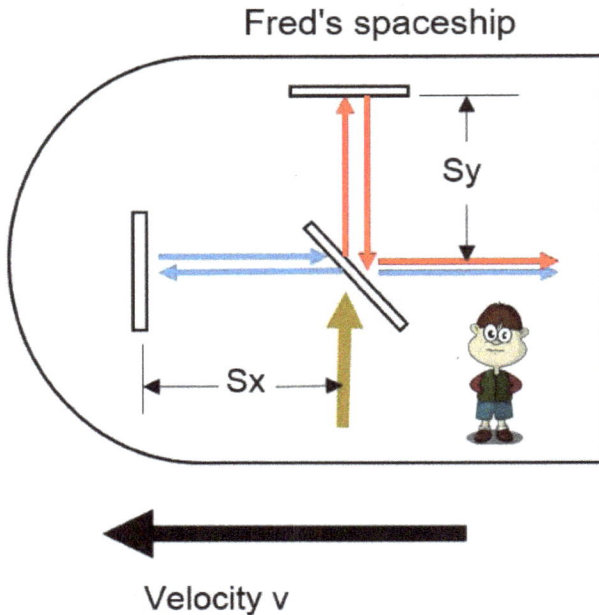

Clipart images from clipart.org

You, observing from Earth

But, this is **not** what **you** observe on Earth. The light beam is moving at velocity **c** relative to you, but Fred's spaceship is moving in the same direction at velocity **v**, so the light beam must be moving at velocity **(c-v)** *relative to the spaceship*. From your observations, the time required for the light beam to travel to the mirror is equal to distance **Sx** divided by the relative velocity **(c-v)**. Similarly, after being reflected from the mirror, the light beam must travel at velocity **(c+v)** relative to the spaceship.

According to *your observations*, the total time required for the light beam to make the round-trip journey is:

$$\text{Time for light beam to travel along x-axis} = \frac{Sx}{c-v} + \frac{Sx}{c+v} = Sx \left[\frac{1}{c-v} + \frac{1}{c+v} \right]$$

To add the two terms within the brackets, we convert these fractions to the lowest common denominator:

$$\text{Time for light beam to travel along x-axis} = Sx \left[\frac{c+v}{(c-v)(c+v)} + \frac{c-v}{(c-v)(c+v)} \right]$$

Simplifying, we get:

$$\text{Time for light beam to travel along x-axis} = Sx \left[\frac{2c}{(c^2 - v^2)} \right]$$

Now consider the light beam travelling in the y direction. According to Fred's observations, the time for the light beam to travel to the mirror and back is simply the round trip distance **2Sy** divided by the velocity of light **c**.

But, again, this is **not** what **you** observe in your "frame of reference" on Earth. This light beam must be moving with the spaceship in the x direction at velocity **v**. But this light is obviously also moving with some velocity in the y direction towards the mirror. Overall, the light beam must have velocity **c** along the hypotenuse of a triangle, as shown here.

So, how fast is the light beam moving in the y-direction towards the mirror? Applying the Pythagorean theorem to the velocity vectors comprising this triangle, **you observe** that the light beam is travelling towards the mirror at velocity $\sqrt{c^2 - v^2}$.

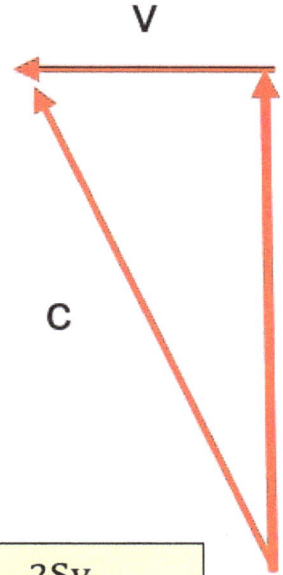

The time for this light beam to travel distance **Sy** to the mirror is equal to **Sy** divided by velocity $\sqrt{c^2 - v^2}$. Exactly the same time will be required for the light beam to undertake the return journey.

Consequently, the total time required for this light beam to complete its round trip journey is:

$$\text{Time for light beam to travel along y-axis} = \frac{2Sy}{\sqrt{c^2 - v^2}}$$

You and Fred both observe that the two light beams arrive exactly in phase, and therefore require exactly the same time to travel on their respective journeys.

So, **from your observations**, the time for the light beam to travel in the x-direction (in the direction of velocity **v**) is equal to the time required for the other light beam to travel in the perpendicular direction. Setting these two terms equal, we get:

$$Sx \left[\frac{2c}{(c^2 - v^2)} \right] = Sy \left[\frac{2}{\sqrt{c^2 - v^2}} \right]$$

Simplifying and re-arranging this equation, we get:

$$Sx = Sy \sqrt{1 - \frac{v^2}{c^2}}$$

From your perspective, distance **Sx** along the direction that the spaceship is moving has shrunk, while distance **Sy** perpendicular to the spaceship velocity remains the same. Not only has distance **Sx** shrunk in the x direction, but you would observe that the entire spaceship and everything in it (including Fred) has shrunk in this dimension. In fact, everything in this "reference frame" (moving at velocity **v** relative to you) has shrunk. That's because – as Einstein reasoned – *space itself has shrunk in this reference frame!*

From your vantage point on Earth, every measurement, every experiment, every observation that you could ever perform would show that Fred's spaceship (and everything moving with it at velocity **v**) has shrunk in the direction that it is moving relative to you. But Fred would detect no such change in the size or shape of his spaceship. With such conflicting viewpoints, you might ask, "what is *really* happening on Fred's spaceship?". This, from the view of special relativity, is a meaningless question. *Reality is different in your reference frame from the reality in another reference frame*.

For Fred, everything in his spaceship (and his reference frame) appears perfectly normal. From his perspective, the Earth is moving away from him at velocity **v**. We cannot distinguish whether the spaceship is moving away from the Earth, or whether the Earth is moving away from the spaceship. No one knows their absolute velocity. All we can say is that the spaceship and Earth are moving at velocity **v** *relative to each other*. Any measurement, observation or experiment that Fred could undertake would show that the Earth and everything moving with it (including you) has shrunk in the direction of relative motion.

But this is not the only difference in reality between your world on Earth and Fred's world on the spaceship. Let's say that you decide to compare units of time in your world and in Fred's world. We can define the unit of time in terms of how long it takes for a light beam to travel any convenient specified distance. Since, both you and Fred each have an identical Michelson-Morley apparatus at hand, you can measure the time for a light beam to travel distance **Sy** on-board Fred's apparatus, and compare that with the time for a light beam to travel the same distance in your laboratory on Earth (alternatively, you could choose distance **Sx**, or any other path, and would get the same result).

As we have seen before, we observe the light beam on Fred's spaceship has a velocity component along the y axis of $\sqrt{c^2 - v^2}$. The time it takes for the light beam to travel distance Sy at velocity $\sqrt{c^2 - v^2}$ is $Sy/\sqrt{c^2 - v^2}$.

But, in our Michelson-Morley apparatus on Earth, we observe that the light beam moves along the y axis at the normal speed of light, **c**. The time required for the light beam to travel distance **Sy** is Sy/c. *From our perspective on Earth*, a light beam takes more time to travel the same distance on Fred's spaceship than in our reference frame on planet Earth. Not only that, but everything that we observe happening on Fred's spaceship occurs more slowly than it does on Earth. From our perspective, Fred's world moves in slow-motion. Clocks move more slowly, people move and talk more slowly, everything takes longer to happen. Once again, this effect is not an optical illusion. Every experiment that you could perform would show that time passes more slowly on Fred's spaceship. We say that "time dilation" occurs.

The number of seconds required for some event to occur on the spaceship (t_{ship}) is related to the number of seconds for the same event to occur on Earth (t_{Earth}) as follows:

$$t_{ship} = \frac{t_{Earth}}{\sqrt{1 - \dfrac{v^2}{c^2}}}$$

If Fred's spaceship is moving at 90% of the speed of light (v/c = 0.9), one second on the spaceship would last about 2-1/2 seconds in "Earth time".

Meanwhile, on-board Fred's spaceship, everything appears to be moving at normal speed. When Fred peers back at Earth, everything seems to be moving more slowly on Earth.

Let's consider this scenario. Imagine that you have an identical twin named George. You and George celebrate your 20th birthday together, before George steps on-board a rocket and blasts off into outer space, travelling at 90% of the speed of light. During the ten-year voyage, you correspond via Skype with your twin, whose speech has slowed to 40% of its previous rate. He goes for months before he needs a haircut. Ten years pass, but George has only aged by 4 years. However, George tells you that **he observes** that **you** are moving slowly and aging less. So who is now older? The only way to find out is to bring you and George back into the same reference frame. For that to happen, one (or both) of you must be accelerated. This goes beyond the realm of the Special Theory of Relatively, which deals with the special case of bodies moving at constant speed. The General Theory of Relativity tells us the answer. If George's spacecraft is decelerated and brought back to Earth, he will be younger than you are.

In principle, we can use "time dilation" to enable us to travel to distant parts of our galaxy within a single lifetime. Let's say that we wish to set off on a rocket to explore a planet discovered orbiting a star that is 1,000 light years from Earth. As a consequence of the Theory of Relativity, we cannot travel faster than the speed of light, so it would take at least 1,000 years for a spaceship (moving near the speed of light) to reach the newly-discovered planet. But that is 1,000 years in Earth time. If the astronauts travel on a rocket travelling at 99% of the speed of light, only 140 years would elapse for the astronauts on board the spaceship. That's much better, but the astronauts would probably still have expired of old age well before they arrive at the destination. If the rocket travelled at 99.99% of the speed of light, only 14 years would have elapsed during the flight. No doubt, the astronauts would be thrilled to report landing on the planet to their family, friends and colleagues back on Earth – but more than 1,000 years would have elapsed on Earth. Even their great great grandchildren, who were born well after they departed on their journey, would have long since passed away.

Of course, to the astronauts on board the rocket hurtling through space, everything would appear normal and happen at its normal pace. But, when they peer at their destination planet through a telescope, it appears to be flattened in the direction that they are travelling. From their perspective, not only is the planet squashed, but so is everything in its reference frame – and this includes the space remaining between the spaceship and the planet. For these intrepid astronauts, the reduced (14 year) travel time is not due to slower passage of time within their reference frame, but to the compression of space through which they are travelling.

15. Special relativity and the equivalence of mass and energy

Previously, we have shown how time and space can expand and contract for observers in different reference frames (moving at different velocities relative to each other). But, we have also seen that, according to Einstein's assumptions, the laws of physics are the same for all observers throughout the universe. One of the most fundamental laws of physics is the conservation of energy.

For you (on Earth) and your cousin Fred (in a spaceship moving at velocity **v** relative to you), space and time are different, but both of you would agree that energy is conserved in your respective reference frames. But what would happen if you transfer energy from your world to Fred's? Let's set up a hypothetical "thought experiment" to see the result.

When Fred's spaceship was launched into space, energy was required to accelerate his spaceship to velocity **v**. The spaceship acquired kinetic energy, which is known from classical physics to be equal to one-half the mass of the spaceship times the square of its velocity (kinetic energy = $\frac{1}{2} m v^2$). Let's say that you wish to further accelerate the spaceship to even higher speed. The way that you do this is irrelevant. For simplicity, we might imagine, although it would be completely impractical, that you push the spacecraft through a long rigid pole. More likely, you could use electrical or magnetic fields, as is done in particle accelerators (like the Large Hadron Collider) to accelerate electrons or protons to speeds approaching the speed of light.

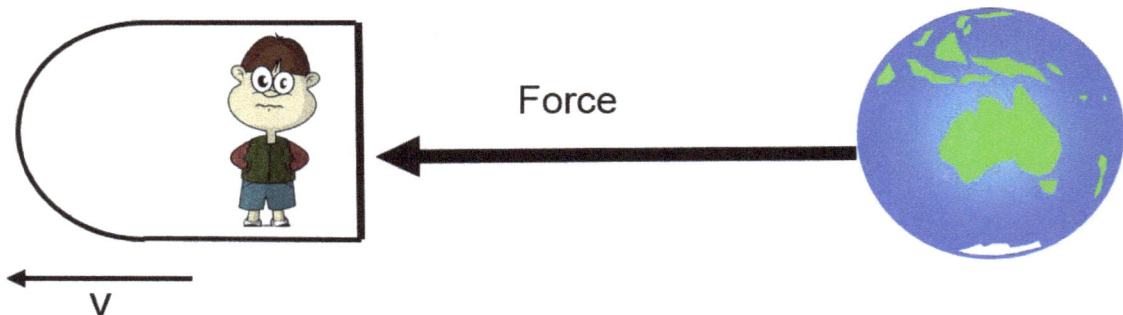

Whatever technique is used, let's imagine that you apply a force to accelerate the spacecraft for a short period of time and to increase slightly its velocity. To simplify the calculations, let's consider the case where a small additional velocity Δv is imparted to Fred's spacecraft (so that the **additional velocity Δv** is much less than **velocity v** that it already has).

By increasing the velocity of Fred's spacecraft by Δv, you are increasing its kinetic energy by **$mv\Delta v$** (where **m** is the mass of the spacecraft and **v** is its original velocity).

The additional velocity Δv gained by the spacecraft varies in proportion with the force applied, and with the time duration over which the force is applied. ***But there's a problem here!*** Let's say that you push on the rod for 10 seconds, and you duly record this in your laboratory notebook. At the same time, your assistant looks through a telescope into Fred's spaceship (which is made of clear plastic, of course), and sees that Fred's watch has advanced only 4 seconds. Your cousin Fred, who is absolutely trustworthy, assures you that only 4 seconds elapsed while you were pushing on his spacecraft. The actual gain in velocity of the spacecraft determined by Fred, Δv_{actual} is less than the velocity that you would have expected $\Delta v_{expected}$ according to classical (non-relativistic) physics.

In fact, your measuring instruments confirm that the spacecraft has only increased in velocity by Δv_{actual}, rather than $\Delta v_{expected}$. Why is this? Let's say that you accelerated Fred's spacecraft by applying a force of 100 Newtons for a distance of 100 metres. The work that you expended in pushing Fred's spacecraft was the force times the distance, or 10,000 Newton-meters (10,000 Joules). At Fred's end of the rod, the spacecraft experienced the same pushing force of 100 Newtons, but the spacecraft was only pushed for a distance of 40 metres (since the spacecraft, and everything else in its reference frame, has shrunk).

From the equations that we derived earlier, we can relate the actual increase in velocity of Fred's spaceship with the velocity change that you expected (based on energy conservation):

$$\Delta v_{expected} = \frac{\Delta v_{actual}}{\sqrt{1 - \frac{v^2}{c^2}}}$$

This means that, while you expended 10,000 Joules to accelerate the spacecraft, it seems that only 4,000 Joules of energy was actually received by the spacecraft. It initially appears that 6,000 Joules of energy somehow got lost in the transfer from one reference frame to another. We had expected the spaceship to gain a kinetic energy of $mv\,\Delta v_{expected}$, but the actual gain in kinetic energy was $mv\,\Delta v_{actual}$. The amount of energy that seems to have gone missing is:

"Missing Energy" $= mv(\Delta v_{expected} - \Delta v_{actual})$

Inserting the equation relating $\Delta v_{expected}$ to Δv_{actual}, we get:

$$\text{"Missing Energy"} = mv\,\Delta v_{actual}\left[\frac{1}{\sqrt{1 - \frac{v^2}{c^2}}} - 1\right]$$

Let's consider the particular situation where the velocity v of the spacecraft is small compared to the velocity of light. This allows an enormous simplification of the maths. As it turns out, the conclusion that we will derive in this particular case is perfectly general, and applies in cases even for very high velocities, approaching the speed of light.

If the ratio v/c is a small number, then the ratio $(v/c)^2$ will be an even smaller number. We can apply the following mathematical simplifications:

For a number n that is much less than 1: $\quad \sqrt{1-n} = 1 - n/2$

$$\frac{1}{1-n} = 1 + n$$

You can test these equations for yourself to see how accurate the results are when $n = 0.1$, $n = 0.01$, $n = 0.001$.

Applying these two simplifications, we find that the "Missing Energy" is:

$$\text{"Missing Energy"} = mv\,\Delta v_{actual}\left[\,1 + \tfrac{1}{2}\frac{v^2}{c^2}\,-1\,\right]$$

Simplifying, we get:

$$\text{"Missing Energy"} = mv\,\Delta v_{actual}\left[\,\tfrac{1}{2}\frac{v^2}{c^2}\,\right]$$

Of course, the "Missing Energy" is not missing. The laws of physics, and the conservation of energy, still apply. So how can we account for the "Missing Energy".

Let's go back to re-consider what has happened. If Fred's spaceship is very light (low mass), it would undergo a large increase in velocity when we push it. If the spaceship were very massive, it would undergo a much reduced gain of velocity. It seems that Fred's spaceship has more mass after we pushed it than it did before. The "extra mass" **Δm** is moving with the spaceship and carries kinetic energy. The kinetic energy of the extra mass accounts for the "missing energy".

We must conclude that, in being accelerated from velocity **v** to velocity (**v+Δv**), the spacecraft had gained mass **Δm**. The kinetic energy of this newly-created mass **Δm** moving at velocity **v** is ½ **Δmv²**. Equating the "Missing Energy" with the kinetic energy carried by the additional mass **Δm**, we get:

$$mv\,\Delta v_{actual}\left[\,\cancel{\tfrac{1}{2}}\frac{\cancel{v^2}}{c^2}\,\right] = \cancel{\tfrac{1}{2}}\,\Delta m\,\cancel{v^2}$$

Re-arranging and solving for the value of Δv_{actual}, we get:

$$\Delta v_{actual} = \frac{\Delta mc^2}{mv}$$

The kinetic energy **ΔE** imparted to the spaceship as it was accelerated from velocity **v** to velocity (**v+Δv**) is:

Gain in energy, $\Delta E = mv\,\Delta v_{actual}$

Substituting for the value of Δv_{actual} derived above, we get:

$$\text{Gain in energy, } \Delta E = \cancel{mv}\left[\frac{\Delta mc^2}{\cancel{mv}}\right]$$

So, $\Delta E = \Delta m\,c^2$

The gain in mass of the spaceship is directly related to the energy imparted to the spaceship.

We have derived this result for the specific case of a spacecraft moving at velocities that are small in relation to the speed of light. However, using the same type of reasoning, Einstein showed that, *as a completely general rule, energy and mass are related according to the equation*:

$$E = m\,c^2$$

This implies that, even when a particle is not moving at all, it still has energy equal to its "rest mass" m_o multiplied by the speed of light squared.

Energy can be converted into mass – and mass into energy – according to the equation $E = mc^2$. Note, however, that a huge amount of energy is required for each additional kilogram of mass. For example, Australia's total annual production of electricity, which would require 60 million tonnes of coal to be burned (equivalent to a stockpile 10 kilometres in diametre and more than 3 kilometres high) could produce an extra 10 kilograms of mass.

The conversion of mass into energy is not a hypothetical, theoretical concept, but has been observed in many situations. It explains why we cannot accelerate a particle faster than the speed of light. If we tried, we would find that the particle becomes more and more massive as its velocity approaches the speed of light. We would need to invest more and more energy to achieve each slight increase in velocity. An infinite amount of energy would be required to bring the particle to the speed of light.

The mass of a particle can be related to its "rest mass" m_o by the equation:

$$\text{Mass } m = \frac{m_0}{\sqrt{1 - \frac{v^2}{c^2}}}$$

Note that, as the velocity of a particle approaches the speed of light, its mass approaches infinity.

The energy of a particle can also be related to its "rest mass" m_o by the equation:

$$\text{Energy} = \frac{m_0 c^2}{\sqrt{1 - \frac{v^2}{c^2}}}$$

In the special case where the velocity of a particle is much less than the speed of light, we can use the simplifying mathematical formulas given earlier to show that:

$$\text{Energy} = m_o c^2 + \tfrac{1}{2} m v^2 \qquad \text{In the case where } v \ll c$$

The first term ($m_o c^2$) gives the "rest energy" of the particle (when it is not moving), and the second term ($1/2\ mv^2$) is the kinetic energy in classical physics.

The objects and particles in our normal day-to-day world move at only a tiny fraction of the speed of light. A bullet from a 9 mm pistol travels at only about one-millionth of the speed of light, and we certainly would not notice (or even be able to measure) its 0.00000000005 % gain in mass. But for particles like electrons, protons and atomic nuclei that are accelerated to high energy, the gain in mass is obvious and unmistakable.

In the early decades of the 20th century, physicists were beginning to discover nuclear reactions, and finally understood that the energy produced by our sun (and by all stars in the "main sequence" phase of their lives) is derived from the fusion of four hydrogen nuclei into one helium nucleus. This is the most energetic nuclear reaction that is known. The mass of four hydrogen atoms is about 0.7% more than the mass of one helium atom, and this mass "disappears" during the fusion reaction. Actually, this mass doesn't disappear. It is converted into heat energy which eventually escapes as radiation from the surface of the sun.

For experiments in the Large Hadron Collider, the largest particle accelerator ever constructed, protons are accelerated to 99.999999% of the speed of light and then collided with other atomic particles. At these speeds, protons have about 7,000 times more mass than at rest, and accordingly, huge amounts of energy are required to acceleratre proton beams to this speed.

We have considered the change in mass of a particle as it gains or loses energy. However, particles can also be **created** from "pure energy". When a high-energy gamma ray interacts with an atomic nucleus, the interaction can produce an electron and a positron (a particle with the same mass as an electron, but with a positive electrical charge). This process was first observed by Patrick Blackett, who received the Nobel Prize in 1948 for his discovery of "electron-positron pair production".

Blackett discovered electron-positron pair production by observing the interaction of cosmic rays (high-energy gamma rays and sub-atomic nuclei, probably produced by supernova explosions) with air molecules in a cloud chamber. One of the first such cloud chambers is still operational and housed at the Exploratorium Science Centre in San Francisco. When I visited the Exploratorium a few years ago, I was amazed to see the tracks of high-energy electrons, positrons and other particles directly in front of my eyes! (you can see it too; have a look at: http://video.mit.edu/watch/cloud-chamber-4058/)

For an electron and positron to be produced in this way, the gamma ray must have an energy greater than the combined rest mass of an electron and a positron. The mass of an electron and positron are known precisely and, according to Einstein's equation $E = mc^2$, for an electron-positron pair to be formed, the energy of the gamma ray must exceed 1.02 million electron volts (the combined mass of an electron and positron

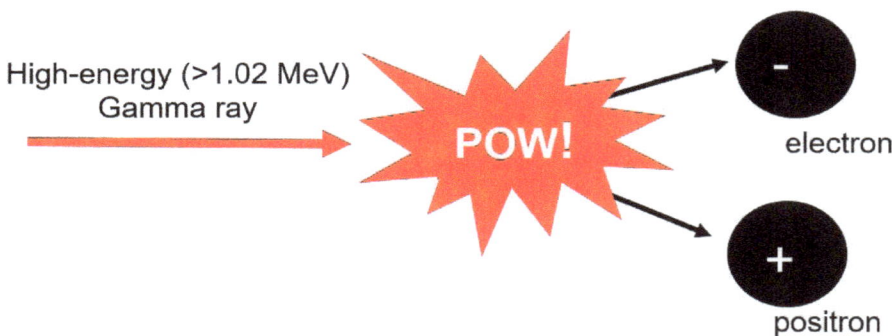

High-energy (>1.02 MeV) Gamma ray

POW!

- electron

+ positron

multiplied by the speed of light square). This is about a million times greater than a photon of visible light. If the gamma ray has greater energy, the surplus is imparted as kinetic energy to the electron and positron.

The opposite process, called "electron-positron" annihilation, also occurs. In this case, electrons and positrons collide and disappear, producing gamma rays with an energy of at least 1.02 million electron volts.

It is believed that electron-positron pair production was a critical step in the creation of matter during the "big bang" which created the universe. In the first seconds after the "big bang", the universe consisted only of gamma ray radiation of unimaginable energy and intensity. This radiation would have undergone electron-positron pair production and similar processes (perhaps creating protons and anti-protons), giving rise to the matter that remains today. However, pair production produces equal numbers of electrons and positrons, so it is a mystery why the current universe consists (we think) almost entirely of electrons, protons and such "matter", rather than positrons, anti-protons and anti-matter.

An animated summary of the Theory of Relativity can be viewed in a 16 minute video:
https://www.youtube.com/watch?v=ev9zrt__lec   (note: double underscore)

16. Electricity, electronics and the digital revolution
Introduction

The 1950s and 1960s was an amazing period to be growing up in New York City. Television was the new consumer craze, overtaking radio as the wonder-technology of my parent's generation. Even families of very modest income, like my own, had a television. Of course, the television pictures were black-and-white, and the picture quality was terrible. Adjusting the position of the antenna (sticking out the window of my parent's bedroom) was a regular ritual, and even then, ghost images and extra lines moving across the screen were routine.

Our television only gave a (barely) watchable image on two channels. Once each week (I think it was Tuesday night), we would visit another family living on the 5th floor of our building and we would all watch the "World of Disney" on their television, which received Channel 7.

My childhood marked the transition from electronics based on glass-and-metal tubular devices called "valves" in the UK and Australia (because they turned on and off electric currents, like a valve) and called "vacuum tubes" or just "tubes" in the US. These were the pinnacle of technology in the 1950s. Electronic valves were intricately constructed using quite impressive science and sophisticated techniques (fine metal electrodes, ultra-high vacuum, impervious glass-metal seals, and cathode surfaces with high electron emission). They had become much smaller than the preceding fist-sized generation of electronic valves, being roughly about the size of a person's thumb. They probably were – and remain to this day – the most sophisticated and elegant technology that you could see how it worked with your unaided eye.

Within a half-hour's walk of where we lived was the area around Canal Street in lower Manhattan, which hosted a wide range of shops selling electronic parts and miscellaneous hardware. These included "army surplus stores" selling every conceivable type of equipment and components – knobs, military radios, portable generators, bombsights – you name it. Everything was laid out on shelves and in open boxes. My brother, father and I spent hours "kibitzing" (examining and drooling over the merchandise in the store, and then usually leaving without buying anything).

It was not unusual for geeky nerds, like my brother and I, to build our own radios and electronic equipment. My brother became interested in "amateur radio", communicating directly with other "ham" radio operators around the world. At that time, it was possible for talented hobbyists to build their own short-wave radio receivers and transmitters that were better than any that you could buy. My brother assembled his own "ham rig" by converting surplus military equipment (he started with an ACR-5 radio receiver, which was standard equipment on US Navy aircraft in WWII). I built a few electronics projects, which often didn't work because I was determined to design the circuits myself. These included a small "super-regenerative" receiver using two miniature "acorn tube" valves, which I liked because they were small and cute, and really cheap (they were designed to be used in radar sets). My radio worked for a short time (my construction and soldering technique was pretty lousy).

One project that I built was a great success. When I was in high school, my friend Alan regularly played the quiz game Jeopardy with his parents, sister and aunt and uncle (who lived in the same building). They were Jeopardy fanatics, playing one or two evenings each week, but were frustrated with the simple noisemaker provided with the game for players to indicate that they knew the answer. For easy questions, all the players lunged at their clickers nearly simultaneously, and it was impossible to tell who responded first. So, they asked if I could build a device that could accurately show who responded first. So, I designed and built a

"Jeopardometer". Each player had a push button which activated a small neon indicator lamp. There was an indicator lamp for each player, but a lamp would only light for the first person to press their switch.

The "Jeopardometer" circuit that I designed, I must say, was brilliantly simple. It was based on a particular characteristic of the neon indicator lamps. About 80 volts was required for a lamp to light, but once turned on, the lamp remained lit with a constant voltage of 60 volts. So, once the first player pressed their switch to light their indicator lamp, there was no longer sufficient voltage to light any other lamps. Alan's family used the Jeopardometer frequently, and I would occasionally get desperate calls for an emergency service visit when one of the wires had come loose or a switch had stopped working. Years later, I built an improved semiconductor version, but by that time, I had moved away and lost contact with Alan and his family.

By the time I finished high school, manufacturers were mass-producing better and better electronic equipment of various kinds, at lower prices. The motivation for electronic enthusiasts to build their own equipment evaporated. Buying components to build your own equipment cost much more than buying a commercial product off-the-shelf, and building your own equipment was fiddley and time-consuming.

The world of "electronic valves", "army surplus stores" and electronic enthusiasts building their own equipment was rapidly disappearing. Electronic valves could not compete with transistors, which were easily a hundred times smaller, used less power and could operate with a small low-voltage battery. Transistors were not much to look at or see, usually being contained within a blob of back plastic with a few wires sticking out. The intricate germanium or silicon structure inside, the bit that makes it work, is microscopic.

I recall a conversation with myself when I was perhaps twelve years old, wondering where electronics would go in the years ahead. I could see that improved transistors and integrated circuit technology would enable radios, televisions and record players to be smaller and perhaps work better. Basically, I expected incremental improvements on the types of technology that were already available in the 1960s. However, I completely missed the "big picture" of what was to come in the decades ahead – that having millions of transistors on a single chip costing a few dollars would enable extraordinary things to be done that had been impossible to do with a handful of transistors or valves. In short, I could not imagine the revolutionary impact of computers and digital technology on every aspect of life in the 21st century.

Initially, transistors were limited in what they could do. They were well-suited to applications involving low-frequency, low voltage and low power. At first, they were used for audio (sound) amplifiers and portable AM radios. It was possible to buy transistors that operated at high frequency, *or* at high voltage *or* at high power, but these were specialised and expensive. However, as transistor technology got better and better, they could operate at high frequency, high voltage *and* high power (in fact, at higher frequencies, higher voltages and higher power than the electronic valves that preceded them).

Soon, complex circuits containing dozens of transistors were made on a single tiny piece of semiconductor contained within a small blob of black plastic. These were "integrated circuits". As technology improved, the number of transistors that could be fabricated on a single dot-sized speck of silicon increased. In 1965, Gordon Moore (co-founder of Intel Corporation) predicted that the number of transistors that could be fabricated on each silicon "chip" would double every two years. His prediction, later called "Moore's Law", turned out to be astoundingly accurate and prescient. Over the following five decades, the complexity of circuits on silicon chips increased exponentially, with the number of transistors on each chip reaching thousands, then millions, and is now is reported to be billions.

Most of the early chips were "analogue" circuits that previously would have been constructed from 10 or 20 transistors and other components. If we consider building a complex electronic device like building a house, integrated circuits are like pre-fabricated modules that are bolted together, rather than using individual bricks and pieces of timber.

Some chips were "operational amplifiers" that could amplify weak signals by millions of times. Others were "comparators", which simply compare two input voltages and give an output signal if one input voltage is larger than the other. That doesn't sound very impressive, but comparators can detect voltage differences of a millionth of a volt, and this is incredibly useful for feedback and control systems. By using a comparator, it is very easy to construct power supplies whose output voltage is regulated to a set, precise value. The output voltage is compared to a precisely-controlled voltage reference and, if the output voltage is even slightly lower than it should be, the comparator sends a signal to a power transistor to increase the current flowing to the output.

Then, there are "timer" chips which produce a voltage pulse at precisely-determined time intervals, which can be adjusted from a millionth of a second up to minutes or hours. The duration of the pulse can also be adjusted.

Many of the newer chips are "digital" circuits. These have "logic gates" which simply compare two input voltage signals which are either "on" (say, +5 volts) or "off" (zero volts). The gate gives an output signal (say, +5 volts) or no output signal (zero volts), depending upon whether one or both inputs are "on" or "off".

Each gate performs a simple (perhaps trivial) operation, but can perform this operation millions or billions of times per second. By putting together millions of such gates within a single CPU chip (Central Processing Unit) of a computer, incredibly complex calculations and control functions can be performed.

To be honest, I don't understand how millions of gates are used to schedule airline flights, assign seat numbers to passengers, and decide exactly how much fuel a plane will need to safely reach its destination. I doubt that there is any one person on Earth who has a good understanding of each step in the process. Some people specialise in designing integrated circuits, and others design computer hardware (putting together integrated circuit chips, various types of memory, power supplies, input signal processing, and outputs). "Information Technology" (IT) specialists write software code that tells the hardware what to do, but they don't need to tell each gate what to do. Software specialists use pre-packaged machine code, written by others, that translate each instruction into thousands of simple steps that "tell" the CPU how to carry out each instruction. Packages of low-level code are packaged into larger blocks of software that allow computer programmers to develop sophisticated programs, without having to think about *how* their instructions are carried out.

Mankind has reached an extraordinary situation in which we have become completely reliant on computer technology that is beyond a full understanding by any one person. Many young people have grown up with tablets, smart phones and the internet, and have extraordinary skill at operating within the cyberworld. But these "tech-savvy" whizzes are surfing across the top layer of a technology that is very deep.

In the following chapters, I consider the very deepest layers of this technology: the basic scientific principles that underpin the revolution in electronics and computers that has occurred within our lifetimes.

17. Overview of the electronics revolution

In 1948, the year that I was born, the transistor was invented at Bell Telephone Laboratories (located perhaps a two-hour drive from my birthplace). So, the transistor and I both came into the world at just about the same time. One would cause and dominate the technological revolution that has been occurring ever since (alas, it was the transistor, not I).

Of course, electronic devices had been developing over the first half of the 20th century. By the end of the Second World War, radio and radar were in wide use, and television was gaining acceptance in households in the United States (and, soon after, elsewhere). In particular television had a huge effect on shaping the culture, values and beliefs of virtually everyone. These devices were based on vacuum tube amplifiers, called "tubes" in the US and "valves" in the UK and Australia. These were a critical component in all electronic circuits. The television set in my childhood household contained about 12 or 15 of these vacuum tubes (which I know, because my father would take the television apart to try to fix it whenever it stopped working).

Transistors offered huge advantages over vacuum tubes. They were perhaps one-hundredth the size, did not require a filament that would get very hot and eventually burn out, and consequently, they used less power. Transistors also operated at low voltage that could be supplied by a small battery, which was much safer and convenient for portable devices, rather than requiring high voltages supplied by the mains.

Initially, transistor technology was rather limited. The first transistors were made from germanium semiconductor. They readily overheated and burned out. They were limited to applications requiring low voltages, low frequency and low power. Vacuum tube manufacturers tried to compete against the new transistors by developing miniature vacuum tubes, which they called "nuvistors". But, even with the limitations of early transistors, it soon became obvious that transistors would completely replace vacuum tubes.

Over time, the technology for making transistors got better and better. Silicon replaced germanium as the semiconductor of choice. Transistors made of silicon were far more resistant to overheating.

When I was perhaps 8 or 10 years old, my brother and I got our first transistor radio. It contained two transistors, and worked well if you were standing in the right place in our high-rise apartment. We were very proud that we owned this state-of-the-art technology.

Over the following years and decades, transistor technology improved by orders of magnitude. When my brother and I first became interested in electronics, we could only buy transistors that operated at frequencies below a few megahertz (fine for audio amplifiers and AM radio, but not for FM or shortwave radio). The transistors had a maximum voltage rating of perhaps 20 or 30 volts, and they could only handle about one watt of power before they burned out.

Now, transistors (such as those in your mobile phone) can operate at frequencies of **thousands** of megahertz. Special-purpose high-voltage transistors can handle **thousands of volts**. High-power transistors (called "hockey pucks", because they are the size and shape of hockey pucks) deliver **millions of watts** of power to drive electric trains.

The actual semiconductor junction at the heart of a transistor is tiny. Most transistors are encapsulated within a small piece of black plastic (typically a few millimetres in diameter) with three electrical leads. Only a small fraction of the volume is the actual operating bit of the transistor. The rest is mainly plastic that protects the operating bit, supports the electrical leads and carries away heat produced by the transistor as it switches electrical currents on and off.

Soon, manufacturers began to put circuits containing several transistors into a single plastic "chip". By the 1960s, it was possible to buy "integrated circuits" containing 10 or 20 transistors that served as high-gain amplifiers, timer circuits, voltage regulating circuits, tuning circuits for radios, etc. Having more transistors allowed more sophisticated circuits that would perform more functions, and perform them better. It greatly reduced the cost of connecting together many components to make complex circuits.

In 1965, Gordon Moore, the Cofounder of Intel, saw the potential for technology to reduce the size of transistors and the capability to pack more transistors into each chip. He predicted that that the number of transistors being incorporated into integrated circuits would double every two years for the foreseeable future. This became known as "Moore's Law", and it turned out to be perhaps the most prescient and stunningly accurate prediction ever made. The number of transistors in integrated circuits has continued to double about every two years (and so has their capacity) for the past 50 years. By 1971, integrated circuits were being produced with 2,300 transistors. By 1990, integrated circuits containing *a million transistors* were being mass produced. Today's latest iPhones contain a processor chip that is reported to contain *2 billion transistors*.

Most integrated circuits are now used for digital processing. Each transistor serves as a simple switch that is turned on or off by a small input voltage. These simple switches are put together into AND or OR circuits that turn on or off depending upon two input signals (similar to what a single nerve cell does). This function is really very simple. Each logic circuit performs an apparently trivial operation. However, putting millions of such logic circuits together (each one able to perform this function millions of times each second) enables incredibly complicated calculations and other functions to be performed.

I remember, when I was perhaps 12 years old, wondering what advances in electronics would occur in the years ahead. I thought that, we might have smaller radios and the like, and it would be handy to have radios and televisions that were half the size of the latest 1965-era models. But once these devices got maybe half the size, I thought, there would be minimal benefit to going even smaller. I thought that electronics might reach a "point of diminishing returns". I completely missed what was going to happen. I could only imagine the possibilities for doubling or tripling the number of transistors. I didn't "get" that, if you could increase the number of transistors by millions of times, you could do all sorts of things that had previously been completely unachievable and unimaginable. It is these sort of things that have changed the world in the past few decades, and continue to transform our economy and society.

Periodically over the years, commentators have suggested that Moore's Law might have run its course, and that technological limitations would slow or halt further advancement in computer technology. Each time, researchers have developed new techniques to overcome these limitations. Once again, some people believe that the exponential growth in the integrated circuit technology might finally be coming to its end. Transistor junctions in the present generation of integrated circuits contain only a few thousand atoms, and there are serious questions whether smaller transistors would function correctly.

Exponential growth in the capability of computer technology has exerted a fundamental impact on our society and economy and - even if the rate of technological advancement slows or stops - computer technology will continue to exert a huge effect on our lives.

So, let's have a look at the fundamental science that has underpinned the electronic revolution.

Since we can't see the flow of electrons in electronic circuits, let's begin with a very similar situation – the flow of fluids like air and water. In our day-to-day lives of turning on water taps, we have become familiar with the concept of "pressure". The pressure of a fluid is a measure of the amount of potential energy carried per unit volume of fluid. In other words, the pressure tells us the amount of useful work that could be done by each litre or millilitre of fluid.

When we turn on the nozzle on a garden hose, potential energy that has been imparted to the water (by pumps at the reservoir) is converted into kinetic energy, as water streams out of the nozzle at high velocity. However, the fact that water has the *potential* to do useful work doesn't necessarily mean that it *will* do useful work. If we only turn the nozzle a little bit, water will dribble out, without doing any mechanical work. In this case, the potential energy in the water is dissipated as turbulence and heat.

The unit of pressure in the International SI system is the Pascal. A fluid with a pressure of one Pascal has the potential to do one Joule of work for each cubic metre of fluid. The Pascal is a very small unit, so pressure is commonly expressed in kilopascals (1,000 Pascals) or megapascals (one million Pascals). As it turns out, atmospheric pressure is just about 100 kilopascals.

Bear in mind that the useful work that can be done by a flowing fluid depends on the *difference* in pressure. Let's say that we are using a nail gun to construct a house or cabinet, relieving us of the effort to swing a hammer. We need compressed air to operate the nail gun, so let's say that we have a tank of compressed air at a pressure of 500 kilopascals. This is an absolute pressure of 5 atmospheres, so there is a pressure difference of 4 atmospheres between the compressed air in the tank and the surrounding atmosphere. The nail gun takes in high-pressure air from the tank, extracts its useful work, and releases the air at atmospheric pressure. The pressure gauge on the compressed air tank measures the pressure of the compressed air *relative to outside air pressure*. The "gauge pressure" is 4 atmospheres (400 kilopascals). As much as 400,000 Joules of workcould be extracted from each cubic metre of air flowing into the nail gun.

Of course, as we use more and more of the compressed air in the tank, its pressure will tend to drop. To maintain pressure in the air tank, we need to provide a continuing supply of compressed air. An air compressor draws in outside air (at one atmosphere pressure), compresses it to five atmospheres pressure, and pushes the compressed air into the storage tank. Compressing the air requires work, which is usually provided by an electric motor which is powered by electricity. The source of electricity might be a distant wind turbine,

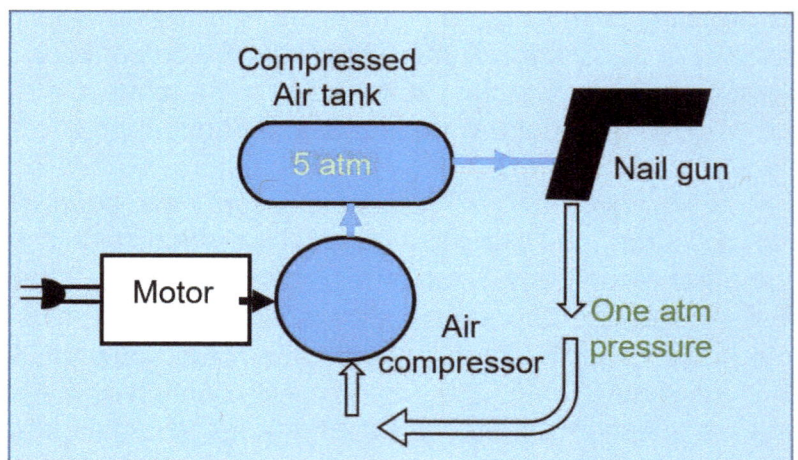

solar panels on your roof or a coal-powered station. So, the kinetic energy in the wind in North Queensland might be providing the work to drive the nails in your nail gun.

In this case, air is used as a medium to carry energy from one source (say, a wind turbine) to a "load" (the nail gun). The air is not changed in the process, and overall, no air is "used up". The amount of air sucked into the air compressor is exactly the same as the amount of air that eventually is exhausted from the nail gun (usually, with an accompanying "kerpow!").

Exactly the same situation would apply to an electric nail gun, or to any electrical circuit.

In recent years, it has become trendy for marketing executives of electricity supply companies to talk about "selling electrons". This irritates me a great deal, because electricity supply companies do not sell electrons. They have never sold a single electron! They simply use electrons as a medium to carry energy from a source (a wind turbine or coal-fired power station) to an appliance you are using. When you plug in a toaster or other appliance, the number of electrons flowing *out* of one terminal on the power outlet is exactly equal to the number of electrons flowing *into* the other terminal.

Electrical circuits are similar to water or air supply systems. In this case, we are concerned with the flow of electrical charge (usually electrons). Charge is expressed in units of coulombs, where one coulomb is the charge of 6.2 billion billion electrons. In flowing into and out of "the load" (say, a toaster), each coulomb of electrons can (potentially) do useful work. The potential energy (Joules) carried by each coulomb of electrons is the "voltage". Thus, "voltage" in electrical circuits is exactly analogous to the "pressure" in water supply systems.

For fluids Pressure (Pascals) = $\dfrac{\text{Potential energy, Joules}}{\text{Volume, m}^3}$

For electricity Voltage (volts) = $\dfrac{\text{Potential energy, Joules}}{\text{Charge, coulombs}}$

In an electrical circuit, we need a "source" to impart energy to the electrons flowing through the circuit. We can use a chemical reaction in a battery to take in electrons that enter via the positive terminal and pump them to the negative terminal (just like an air compressor takes air from the atmosphere and pushes it into a high-pressure tank). Alternatively, we can use an electromechanical generator to utilise mechanical work (produced by a wind turbine, a spinning steam turbine in a coal-fired power station, or the engine of your car) to push electrons through a circuit.

The amount of useful work that can be done in the "load" is determined by the difference in voltage. Accordingly, to measure the voltage difference produced by the source, we place a voltmeter across its two terminals (in parallel with the load). To measure the electrical current (the number of coulombs) flowing *through* the load, we insert an ammeter into the circuit (in series with the load).

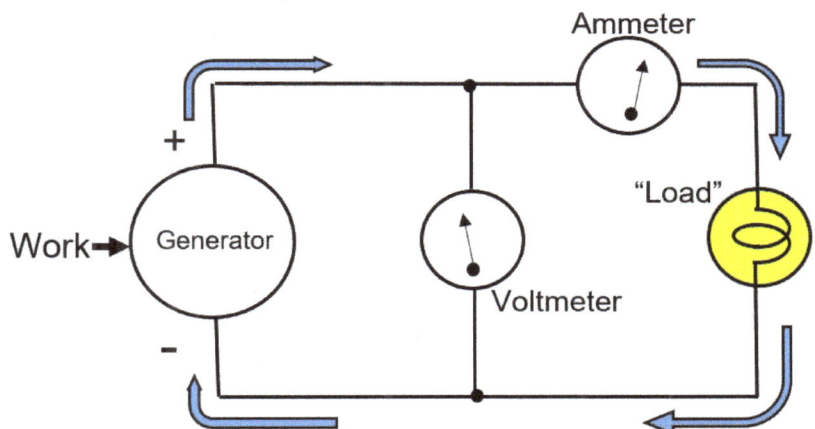

A good way to illustrate how electrical circuits work is to consider the electrical system of a car, which is surprisingly similar to our nail gun example.

But first, I should explain one complication. When electricity was first discovered, scientists did not know about electrons. When they connected a battery (say, a zinc-carbon "dry cell battery") to a load, they observed that something happened, but they didn't know whether electrical charge was flowing from the zinc terminal of the battery to the carbon terminal or vice verse. So, they guessed. They labelled the carbon electrode as the positive (+) terminal, and the zinc electrode as the negative (-) terminal. They had 50-50 odds of getting it right, and . . . they got it wrong. And they didn't know that they got it wrong until decades later, when the electron was discovered and its charge measured. Consequently, electrical engineers still use the convention that electrical charge flows from positive-to-negative, while electrochemists know that it really flows in the opposite direction. To keep things simple, let's adopt the convention that is universally used in electrical engineering, and consider that electrical current flows from positive to negative. The fact that it doesn't turns out to be irrelevant, providing that we always use the same (wrong) convention.

The modern automobile contains a plethora of electrical devices. Just for a start, there is the starter motor, headlights, tail lights, brake lights, interior lights, adjustable side-view mirrors, electric door locks, windscreen wipers, air conditioner fans, and radio/CD player. Don't forget that modern cars also contain an electronic "Engine Management System" that control all aspects of engine operation (fuel injection, ignition timing, etc). Some cars have electrically-adjustable seats and sun roofs. So, the car electrical system has many electrical loads, all connected in parallel.

Each car has a battery (equivalent to the compressed air tank) that provides power to start the engine and to operate appliances while the engine is off. Once the engine is running, an electrical generator backs up and recharges the battery. A regulator maintains the voltage at the correct level (regardless of how fast the engine is going) and prevents the battery from over-charging.

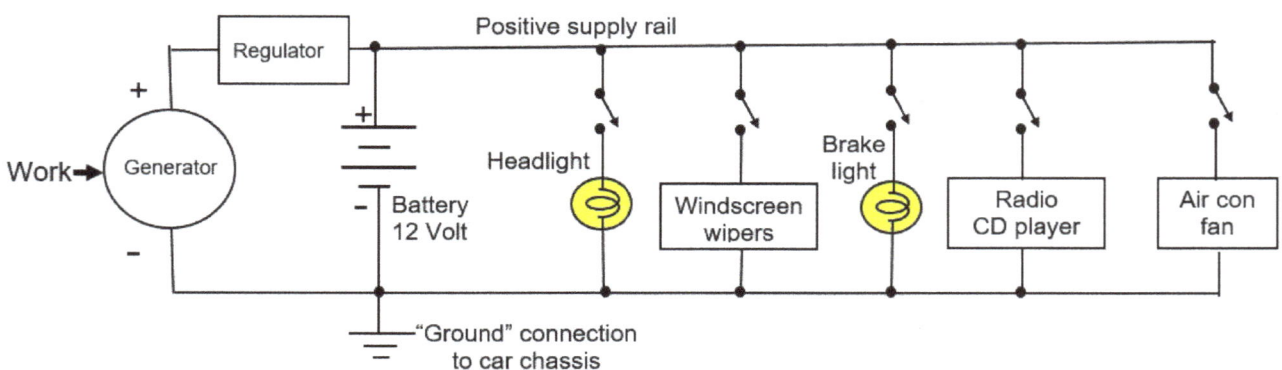

Since cars have rubber tyres, the electrical system of a car is electrically isolated from the outside world. There is no connection to the Earth. Modern cars have electrical systems operating at 12 volts – that is, with a difference of twelve volts between the terminals of the source. By convention, the negative terminal of the battery and generator is connected to the steel car chassis and body. This is the reference voltage – just like all pressures in a nail gun system are relative to atmospheric pressure. The car body is used as a return path for electrical charge to get back to the battery and generator. We say that the car chassis is at "ground potential", or zero voltage (in terms of the isolated world of the car, which is not necessarily at the same potential as the actual ground). Then, each device (headlight or whatever) needs only one wire from its control switch, halving the number of wires running all around the vehicle. Electrical power is tapped from a cable connected to the other (positive)

terminal of the source. This cable, called the "positive supply rail" (or "positive bus") provides electrical power at +12 volts to all the loads in the vehicle.

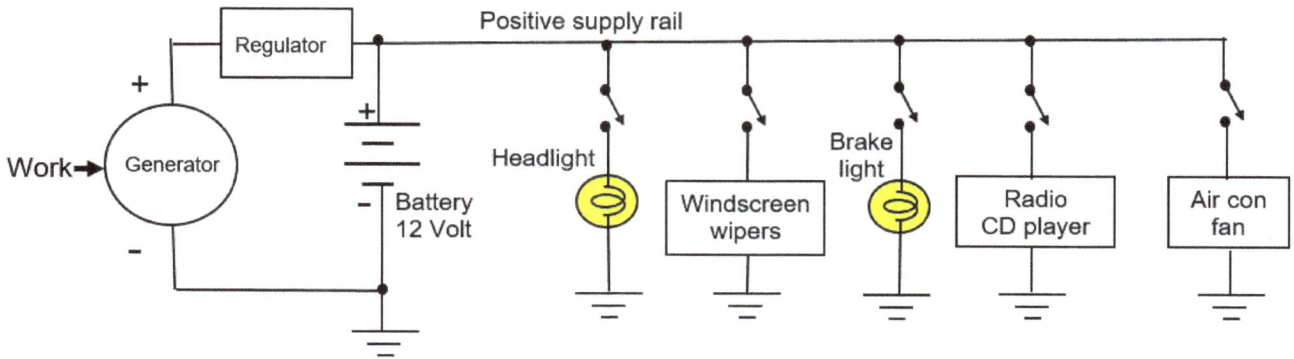

The same configuration is used in virtually all electrical power systems and electronic circuits.

18. Electrical conductivity and resistance to flow

The previous chapter discussed how the voltage difference across a load is the energy expended per coulomb of electrical charge flowing through the circuit. We saw how this is exactly analogous to the role of pressure (energy expended per volume of fluid) for air or water flowing through pipes.

Now, we need to consider how much electrical charge flows through a circuit each second. The unit of electrical charge is the coulomb (about 6 billion billion electrons), and the rate that charge flows through a circuit is given by the number of coulombs per second. One coulomb flowing per second is one ampere. Keep in mind that electrical charge (like water) flows *through* a circuit. For each electron that flows out of the negative terminal of a battery, one electron flows into the positive terminal.

Just as a pressure difference causes fluid to flow, a voltage difference causes electrical current to flow through a circuit.

Imagine that we connect an air compressor to a long thin pipe. The compressor produces a difference in air pressure, and this pushes air through the pipe. High pressure is produced at the outlet of the air compressor (as indicated by dark blue shading), but the pressure reduces continuously as air flows along the narrow pipe until it eventually returns to the compressor (or to the atmosphere).

In this case, high pressure air entering the pipe has the potential to do work, but in fact, no useful work is done. Energy imparted to the flowing air is dissipated as turbulence, and simply causes the air (and the pipe) to get hot.

My reason for describing such a wasteful and pointless experiment is to illustrate some simple concepts. Note that air is pushed through the pipe by the pressure difference produced by the air compressor. The **volume of air** flowing each second through the pipe varies directly with the pressure difference across the pipe. Furthermore, the flow of air increases if we use a pipe with a larger cross-sectional area, and the flow decreases if we use a longer pipe. We can express this with an equation that looks like this:

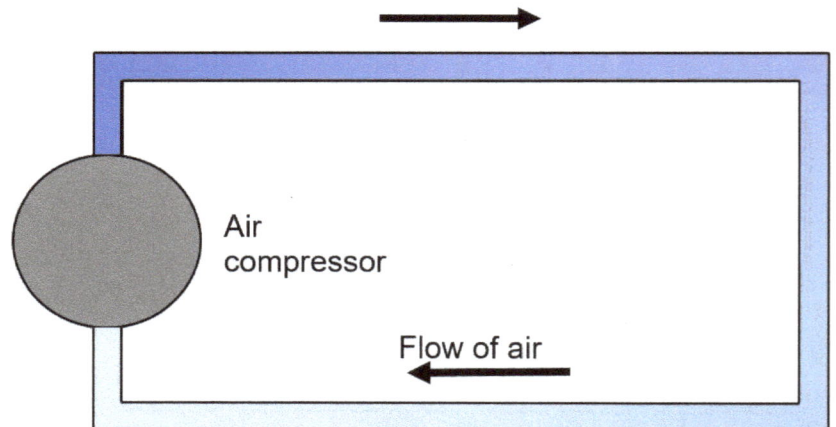

Equation (1)   Flow rate (m³/second) $= P\dfrac{KA}{L}$

Where
- **P** is the pressure difference pushing air through the pipe
- **A** is the cross-sectional area of the pipe
- **L** is the length of the pipe
- **K** is a finagle constant (having something to do with the viscosity of air and the roughness of the inside surface of the pipe).

Now let's construct an analogous electrical circuit. Let's say that we use a battery to produce an electrical voltage difference **V**, and that we apply this voltage across a slab of material. It should seem intuitively reasonable that the amount of electrical current flowing through the slab varies directly with the voltage difference **V** applied across the slab. The electrical current also increases with the cross-sectional area **A** of the slab (since the electrical current has more paths to flow through the material), and varies inversely with its length **L**.

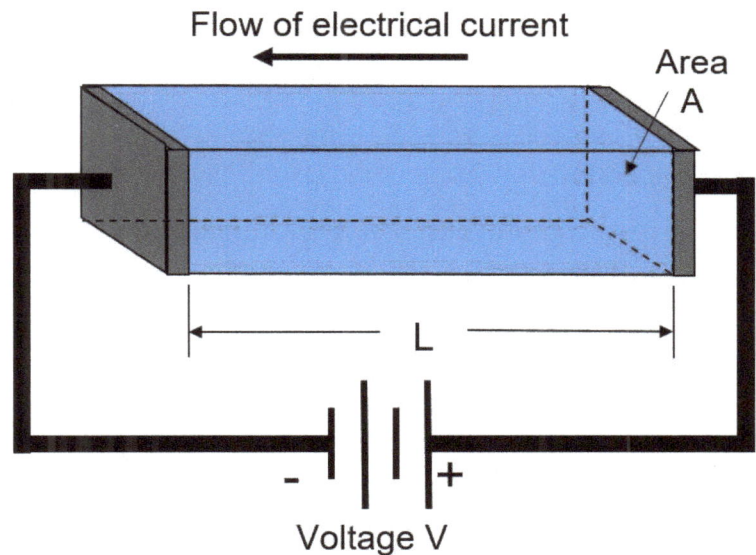

Once again, we can write a simple equation relating the voltage difference and current flow.

Equation (2)   Electrical current (coulombs/second) $= V \dfrac{K\,A}{L}$

Here, the constant **K** relates the ability of the particular material in the slab to conduct electricity. **K** is called the "specific conductivity" of the material. The specific conductivity depends on the temperature, but usually, not by much.

Rather than talk about the "conductivity" of a material, it is generally more useful to talk about its "resistance" to the flow of current. We can rewrite Equation (2) in the form:

Equation (3)   Electrical current (coulombs/second) $= \dfrac{V}{R}$

Where **R** is the resistance of the slab of material, which depends upon its cross-sectional area, length and the "specific resistivity" of the material **σ**. The "specific resistivity" of a material is the inverse of its "specific conductivity"). The unit of resistance is the "ohm" (whose symbol is Ω).

$$\text{Resistance, } R = \dfrac{\sigma\,L}{A}$$

Equation (3) is termed "Ohm's Law" and is one of the foundations of electrical engineering and design. It is almost impossible to design any electrical circuit without using Ohm's Law.

Usually, when we construct electrical circuits, we want to use wires and cables that are good conductors of electricity. Typically, wires and cable are made with conductors of copper (or sometimes, aluminium). Depending upon the amount of current to be carried by the cable, and the length of the cable, we need to use wires that are sufficiently thick (sufficient cross-sectional area). In order for electrical cables to be flexible, the conductors often consist of many thin strands of copper wire. Furthermore, we want to sheath the copper wires in a sleeve of insulating material that doesn't conduct electricity – that is, a material that has extremely low "specific conductivity" and therefore has extremely high "specific resistance".

Generally, there are two types of materials in the world – those that are really, really good conductors of electricity, and those that are really, really bad conductors (really good insulators):

- Metals are typically quite good conductors of electricity. Copper is one of the best conductors, with a specific conductivity of 60 million Siemens/metre. Steel is not nearly as good a conductor. Its specific conductivity is one-tenth that of copper.

- Most other materials (plastics, ceramics, glass, rubber) are extremely poor conductors of electricity (and hence, are good insulators). Glass has a specific conductivity of about one-billionth of a Siemen/metre, about the same as diamond. Plastics like PET and teflon have specific conductivities that are 100 million times less than glass.

This means that even a "poor conductor" like steel conducts electricity 10 billion billion times better than "poor insulator" like glass. Consequently, virtually all materials cam be considered to be either "conductors" or "insulators".

There are a few materials whose specific conductivity lie somewhere in the middle of the huge gulf between "conductors" and "insulators". You might be surprised that water falls in this category. Ultrapure (deionized) water is actually a pretty good insulator, but if salts are present, its conductivity increases a millionfold. That's why it's considered a bad idea to have appliances operating on mains-power near bathroom sinks or bathtubs.

So let's consider an example of how the resistance to flow can have significance in our everyday lives. Many of us live, or have lived, in an older house, built at a time when thinner (half inch) pipes were installed to transport water from the water mains in the street to the kitchen, bathroom and laundry taps. These pipes often wend their way for 20 or 30 metres, under the front yard and up, over and along the floor or roof beams. In many cases, scale deposited along the inside wall of old pipes reduces the area through which water can flow. As a result, these long, thin pipes provide significant resistance to the water flow.

Imagine that you enjoying a nice hot shower, lathering ample suds of shampoo through your hair. The shower provides a steady stream of water, flowing at 10 litres per minute. Let's say that the pressure in the water mains under the street outside your house has a pressure of 5 atmospheres (500 kilopascals) above atmospheric pressure. The 10 litres/minute of water flowing from the shower rose causes a "pressure drop" along the pipe, reducing the pressure at the shower rose to 4 atmospheres. That's fine, as your shower rose gives a nice spray at this pressure.

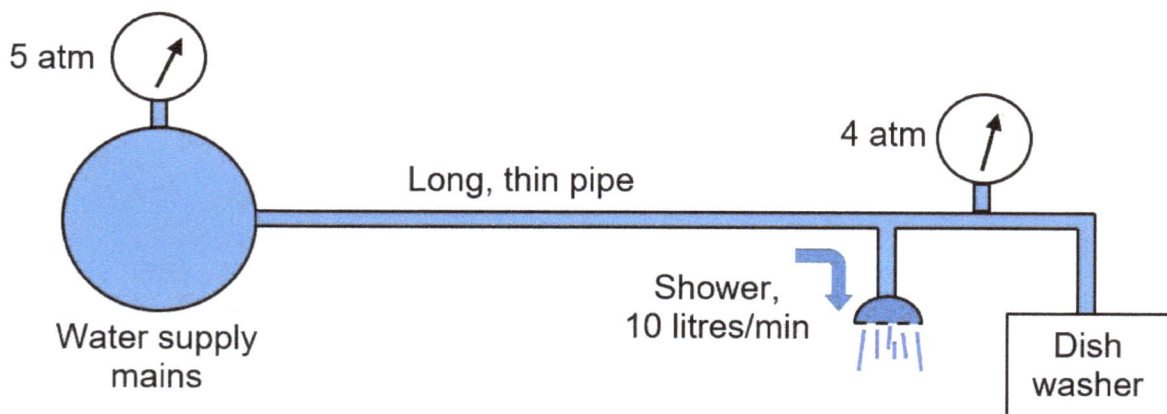

5 atm

4 atm

Long, thin pipe

Shower,
10 litres/min

Water supply
mains

Dish
washer

But then, your spouse, son-in-law or other ingrate decides to turn on the dishwasher at this moment. The total water volume flowing through the supply pipe (including say 25 litres into the dishwasher and a reduced flow of 5 litres/minute to the shower) is now three times as much as before. The pressure drop in the pipe is three times as great, so that the pressure at the shower rose is now only 2 atmospheres. You are left with a dribble to wash the shampoo suds away.

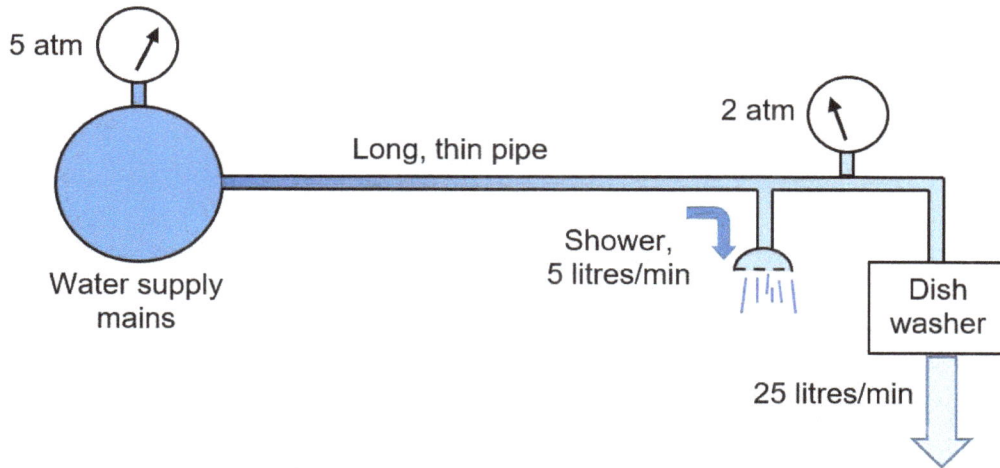

In the same way, electrical cables carrying electricity from the switchboard of a house to various power outlets may be 20 or 30 metres in length. Particularly in older houses, electrical cables may not be of the same standard used in modern houses. Thirty or fifty years ago, no one envisaged the vast array of electrical devices and appliances that are now used on a day-to-day basis. When I first arrived in Brisbane, hardly any houses had air-conditioning; now it is very common. Consequently, when high-power appliances are used, there may be significant voltage drop caused by the resistance in under-sized supply cables.

Let's say that you are sitting in your easy chair one evening, enjoying the tranquillity of a quiet evening as you read these notes. A few lamps, perhaps using one ampere of electrical current, are the only electrical devices in use.

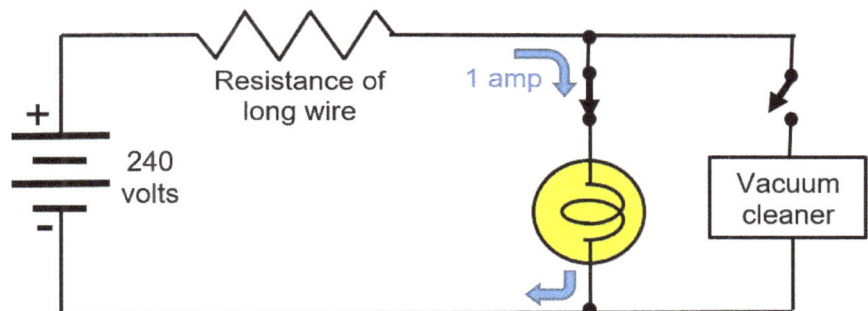

Suddenly, your son-in-law, trying to impress you with his fastidious tidiness, turns on the vacuum cleaner. Electrical motors often draw very large electrical currents when first turned on (usually, this current surge lasts only about a second or less)(Note 1). Let's say that the vacuum cleaner motor uses 9 amperes of current. A total current of 10 amperes flows through the mains power cable, and this would cause the voltage at the end of the cable to fall. Although the voltage at the switchboard might be 240 volts, the voltage at your reading lamp might drop down to 200 volts during the momentary period when the vacuum cleaner motor is starting. This would cause the reading lamp to dim noticeably.

We can apply Ohm's Law to find how much resistance is in the power cable from the switchboard. The "voltage drop" **V** across the cable is 40 volts, the current **I** through the cable is 10 amperes, so the resistance **R** of the cable is given by:

$$\text{Electrical current } \mathbf{I} = \frac{\mathbf{V}}{\mathbf{R}}$$

Re-arranging the Ohm's Law Equation gives:

$$R = \frac{V}{I} = \frac{40 \text{ volts}}{10 \text{ amperes}} = 4 \text{ ohms}$$

Normally, electrical circuits are designed so that the resistance of connecting wires is insignificant. However, electrical circuit designers often use components called "resistors" to deliberately cause a voltage drop and to limit the electrical current flowing through the circuit. Resistors are widely used in electrical circuits. They are available with resistances ranging from a fraction of an ohm to millions of ohms.

A 1,600 ohm resistor. The resistance is indicated by the coloured bands (brown = 1: first digit, green = 6, second digit, orange indicates 3 zeros). The gold band indicates +/- 5% accuracy. The power dissipation rating is ½ watt.

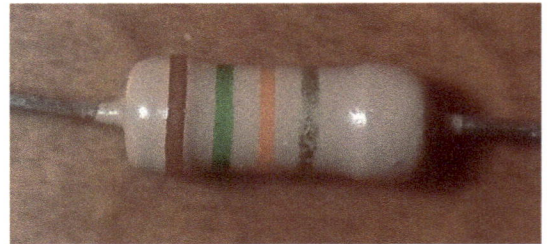

For a simple example of a voltage divider, consider the circuit below:

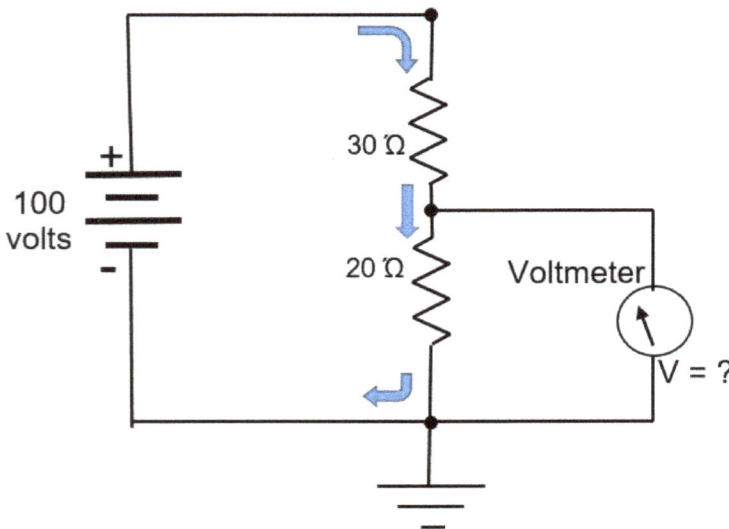

Electrical current from a 100 volt battery flows through a 30 ohm resistor and a 20 ohm resistor in series. The total resistance is 50 ohms, so from Ohm's Law, the current flowing through the two resistors is:

$$I = \frac{100 \text{ volts}}{50 \text{ ohms}} = 2 \text{ amperes}$$

By re-arranging Ohm's Law, we see that the "voltage drop" across the 30 ohm resistor is equal to the current flowing through the resistor (2 amperes) multiplied by its resistance (30 ohms):

$$V = I R = (2 \text{ amperes})(30 \text{ ohms}) = 60 \text{ volts}$$

In the same way, we can determine that the "voltage drop" across the 20 ohm resistor is 40 volts.

This is exactly what we might intuitively expect:

- The voltage drop across each resistor is in proportion with its resistance:

- The sum of the voltage drop across each resistor is equal to the power supply voltage.

Resistors convert useful electrical energy into heat, and the resistors must be able to dissipate this heat fast enough to avoid overheating. In the case of the 20 ohm resistor, a voltage drop of 40 volts appears across the resistor. This means that 40 Joules of energy are released for each coulomb of electrical charge flowing through the resistor. The electrical current through the resistor is 2 amperes, or 2 coulombs per second. So, the total energy dissipated as heat within the resistor is (40 Joules/coulomb)(2 coulombs/second), or 80 Joules/second. This is 80 watts, so the resistor would need to have a power rating of at least 80 watts.

Notes
1. Increasingly, appliances with electric motors use electronic motor controllers to gradually bring the motor to operating speed without a large initial surge in current. The motor on my swimming pool pump employs such a "soft start" feature.

19. Semiconductors

Metals are good conductors of electricity. Generally, metal atoms have one or two electrons in their outer shell, and within the crystal structure of a metal, these electrons are not strongly bound to their parent atoms. Rather, the electrons are free to migrate through the structure. When an electric field is applied, unbound electrons can drift through the metal, transporting electrical charge. It is these unbound, mobile electrons that are responsible for the high electrical conductivity of metals.

In insulators, on the other hand, outer shell electrons of the atoms are tightly bound within molecular orbitals that bind the atoms together. Even if a strong electric field is applied, these electrons are confined within their localised orbitals, and cannot migrate and carry electrical charge.

Semiconductors, like silicon, have particular properties that make them distinct from conductors and insulators. Their electrical conductivity is billions of times less than metal conductors, but billions of times more than common insulators.

Consider silicon (chemical symbol **Si**), which is the most widely used semiconductor material. Atoms of silicon have four electrons in their outer shell. Such a half-full outer shell is unstable, so to achieve a lower state of energy, each silicon atom shares it outer shell electrons with four neighbouring silicon atoms. Although this is not as favourable as actually having a full outer shell, sharing of electrons allows each silicon atom to "feel like" it has a full outer shell (with eight electrons).

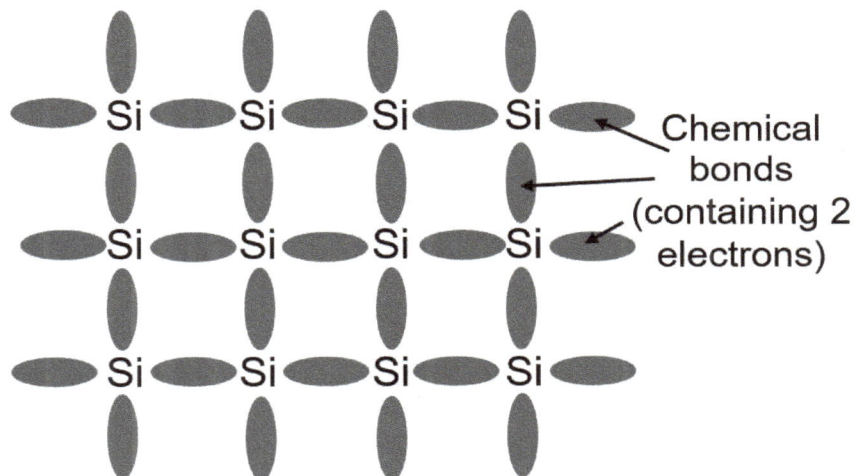

Each silicon atom forms four covalent bonds, sharing each of its outer shell electrons with another silicon atom. Crystals of silicon are three-dimensional structures, but we can represent this in two dimensions as shown here.

With the electrons held within localised chemical bonds, it might appear that silicon should be a good insulator. However, the covalent bond between neighbouring silicon atoms is not that strong. At room temperature, atoms and electrons in a silicon crystal have "thermal" (heat) energy which they exchange with each other. Their average energy is proportional to the absolute (Kelvin) temperature of the crystal which, at room temperature, is about forty times less than the energy binding the electrons within the covalent bonds of a silicon crystal (1.1 electron volts). Nonetheless, heat energy is not uniformly distributed amongst the atoms and electrons. A tiny, tiny fraction of electrons (about one in a billion billion) have sufficient energy to break free of the covalent bond holding them within their localised molecular orbital. These electrons are then free to wander through the crystal.

89

free electron ●   "hole" ⬭

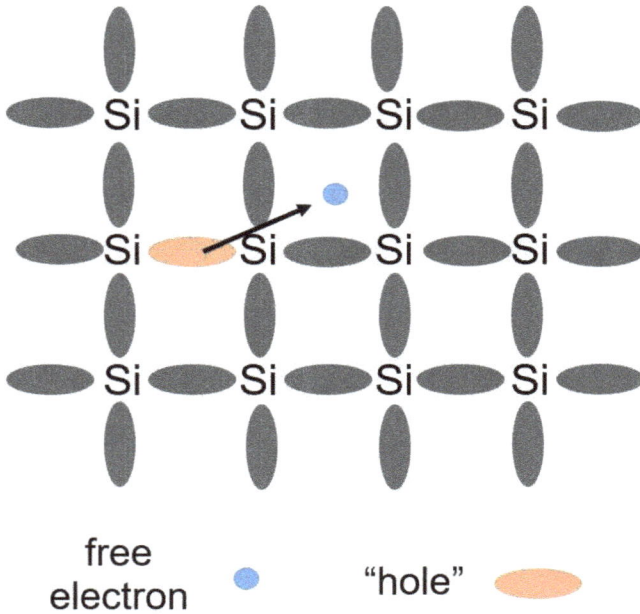

The escape of an electron leaves a molecular orbital with a missing electron. This is unstable, as the two silicon atoms no longer "feel" like they have a full outer shell of electrons. Deprived of their full complement of electrons, the two silicon atoms can seize an electron from a neighbouring molecular orbital. These atoms can then, in turn, seize an electron from other atoms further along the crystal. Of course, this theft doesn't solve the "problem" of the missing electron, it simply passes the "problem" down the line. However, if an electric field is applied to the crystal, the electron is transferred from atom to atom, moving towards the positive end of the electric field. The missing electron, or "hole", moves freely through the crystal structure, exactly like a free electron, but in the opposite direction.

Thus, in pure silicon, there are two types of mobile charges that can move freely through the crystal structure and carry an electrical current – electrons (negatively charged) and "holes" (positively charged).

Bear in mind that, when an electron escapes from the molecular orbital between two silicon atoms, its freedom is only temporary. As electrons wander freely through the crystal structure, they will eventually encounter and recombine with "holes" (which are also wandering around the crystal structure). Within a short time, an equilibrium situation is achieved, where electrons and holes are being formed at the same rate as they are recombining. We can write this equilibrium as:

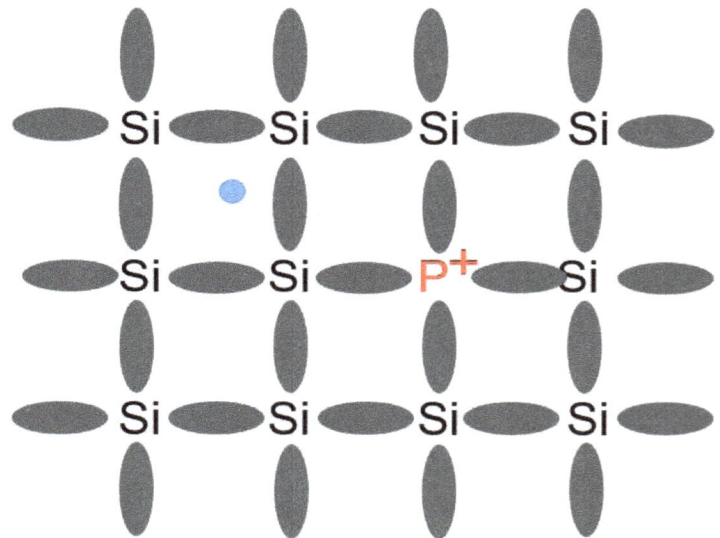

$$\text{Silicon atom} \rightleftharpoons \text{Si+ (hole)} + \text{electron}$$

The two arrows in opposite directions indicate that the process is reversible. Furthermore, the larger arrow in the reverse direction indicates that, at equilibrium, there are many more silicon atoms than electrons and holes. In fact, at equilibrium in silicon at room temperature, there are about 10^{10} electrons per cubic centimetre (and the same number of "holes"). This may sound like a lot, but the number of free electrons and "holes" is about a trillion times less than the number of silicon atoms.

Consequently, pure silicon is a poor conductor of electricity. However, addition of tiny traces of certain impurities (at a concentration of about one part-per-million) dramatically affects the conductivity of silicon.

Consider the effect of adding a tiny amount of phosphorus. Atoms of phosphorus are roughly the same size and readily substitute for silicon in the crystal structure, but phosphorus atoms have five electrons in their outer shell. Four of their outer shell electrons form covalent covalent bonds with the four surrounding silicon atoms, leaving the fifth electron free to wander through the silicon crystal. Thus, addition of tiny amounts of phosphorus will greatly increase the number of free electrons in the crystal and its electrical conductivity. The crystal is said to be "doped" with phosphorus, and to be an "**n-type**" semiconductor, having a preponderance of **negative** charge carriers (electrons) that are free to wander through the crystal.

Substitution of phosphorus atoms (P) for silicon forms an n-type semiconductor. Each phosphorus atom contributes an electron, which is free to wander through the semiconductor crystal. A positive charge remains with each phosphorus atom, which is fixed in position within the crystal structure.

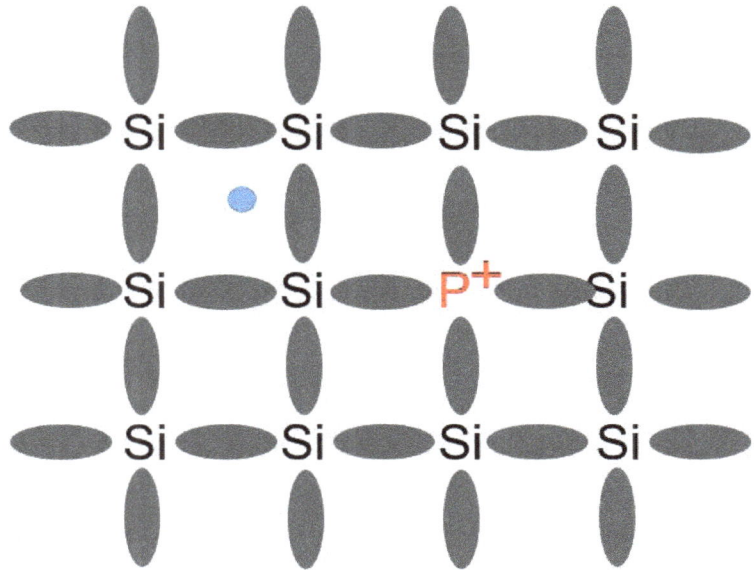

Bear in mind that the crystal is electrically neutral overall. For every free electron contributed by phosphorus atoms, there is a phosphorus atom that has lost an electron and retains a positive charge. Unlike the mobile positive charge of a "hole", this positive charge is locked in position within the crystal structure and unable to move.

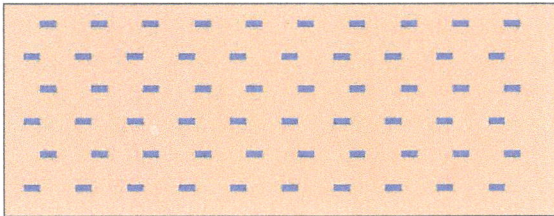

n-type semiconductor
Mobile negative charges (electrons) can move within a matrix of **fixed positive charges**. Positive and negative charges balance, so the crystal has no net electrical charge.

We can represent an n-type semiconductor as follows, showing mobile free electrons in blue. These electrons can wander freely through the semiconductor crystal and conduct an electric current. The same number of positive charges, shown in red, are uniformly distributed through the crystal structure, located on phosphorus atoms bound within the crystal structure.

You might be surprised that, even in n-type semiconductor, there are still some positive "holes" that are free to wander through the crystal structure. But, unlike in pure silicon, the number of "holes" is much less than the number of free electrons. This is due to the equilibrium reaction that we have encountered before:

$$\text{Silicon atom} \rightleftharpoons \text{Si+ (hole)} + \text{electron}$$

In n-type semiconductor, there are perhaps *a million times as many free electrons* as in ultra-pure silicon, so these readily recombine and neutralise "holes". As a result, there are about *a million times fewer "holes"* in n-type silicon as there are in ultrapure silicon.

At equilibrium, electrons and holes are formed at the same rate as they recombine. As a consequence, **the number of free electrons per cubic centimetre multiplied by the number of holes per cubic centimetre is always the same** at a given temperature. For pure silicon at 25°C, the number of free electrons per cubic centimetre Ne_o is about 10^{10}, and this is exactly the same as the number of holes. If we add phosphorus, this dramatically increases the number of free electrons, but also dramatically reduces the number of holes, so that their product is the same as it is in pure silicon.

(Number of free electrons/cm³) (Number of holes/cm³) = (Ne_o) (Ne_o)

$$(N_e)(N_h) = Ne_o^2$$

Where N_e is the number of electrons per cubic centimetre in a silicon semiconductor.

N_h is the number of holes per cubic centimetre in a silicon semiconductor.

Ne_o is the number of free electrons (and the number of holes) per cubic centimetre in pure silicon at 25°C.

Ultrapure silicon has equal numbers of free electrons and holes. By "doping" the silicon with phosphorus to make n-type semiconductors, we can **increase the number of electrons** by a million-fold, but **reduce the number of holes** a million-fold. So, in n-type semiconductor, there are far, far more electrons than holes. We say that, in n-type semiconductors, electrons are the "majority conductors" and holes are the "minority conductors".

It is also possible to do exactly the opposite: to increase the number of holes by a million-fold. This can be done by "doping" pure silicon with a small amount of boron. Atoms of boron have three electrons in their outer shell. When an atom of boron becomes incorporated into a crystal of silicon, three of its outer-shell electrons form chemical bonds with three neighbouring silicon atoms. However, the atom of boron lacks an electron to bond with the fourth neighbouring silicon atom. The chemical bond between the boron atom and this silicon atom will be "missing" one electron. This "hole" can seize an electron from a neighbouring silicon atom and move from atom-to-atom in a "pass the parcel fashion". In the presence of an electric field, the "hole" migrates towards a negative charge.

Thus, adding a tiny amount of boron (about one boron atom to every million of silicon) creates "p-type semiconductor". This contains positive charges that are free to wander through the crystal and conduct electricity, leaving electrons bound to boron atoms distributed uniformly throughout the crystal structure.

Substitution of boron atoms (B) for silicon forms a p-type semiconductor. Chemical bonds to each boron atom lack an electron, which can be captured from nearby atoms in the semiconductor crystal. The resulting positively-charged "hole" can move freely through the crystal. Each boron atom holds an extra electron, giving it a negative charge which remains fixed in position within the crystal structure.

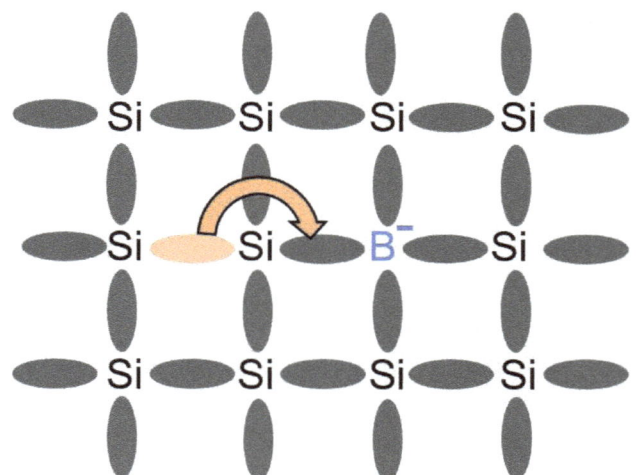

We can represent an p-type semiconductor as follows, showing the mobile positively-charged holes in red. These holes can wander freely through the semiconductor crystal and conduct an electric current. The same number of negative charges, shown in blue, are uniformly distributed through the crystal structure – located on boron atoms that are "locked" in position within the crystal.

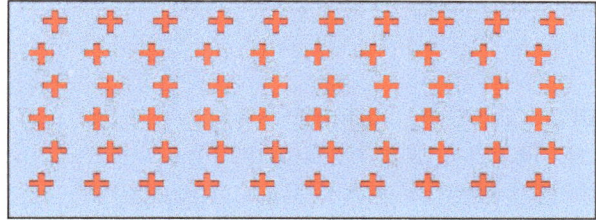

p-type semiconductor
Mobile positive charges (holes) can move within matrix of **fixed negative charges.** Positive and negative charges balance, so the crystal has no net electrical charge.

N-type and p-type semiconductors are the two building blocks of our computer-dominated world. These materials are, from an electrical perspective, "mirror images":

- N-type silicon contains negatively-charged electrons which are free to move through the crystal and transport electrical currents. Electrons are the "majority charge carriers", although an n-type semiconductor also contains far, far smaller concentrations of holes (the "minority charge carriers"). The negative charge of the mobile electrons is balanced by fixed positive charges that are distributed throughout the crystal structure.

- P-type silicon contains positively-charged "holes" which are free to move through the crystal and transport electrical currents. Holes are the "majority charge carriers", although p-type semiconductor also contain far, far smaller concentrations of electrons (the "minority charge carriers"). The positive charge of the mobile "holes" is balanced by fixed negative charges that are distributed throughout the crystal structure.

By joining together a piece of n-type semiconductor and a piece of p-type semiconductor, we can construct one of the useful and important devices in electronics: a p-n junction. As we shall see, a p-n junction allows electrical current to flow in one direction only, and thus serves as a one-way current device, called a "diode". P-N junction diodes are the basis of solar photovoltaic panels (used to generate electricity), LEDs (light-emitting diodes, which are becoming the most common light source for household and commercial lighting, as well as television, computer and mobile phone screens), thermoelectric generators (used on nuclear-powered spacecraft) and LED lasers (used in bar code scanners and DVD players).

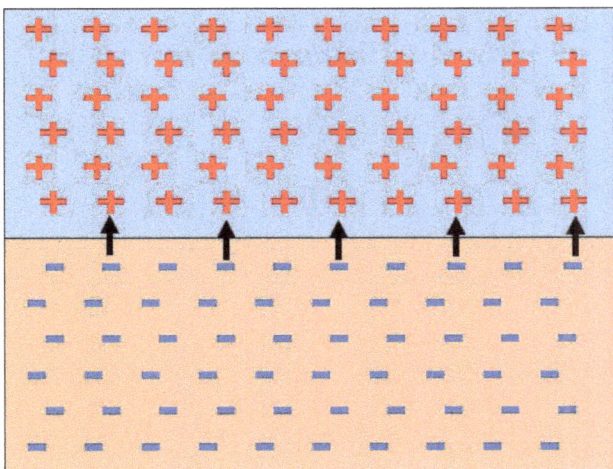

When diodes are made by fusing a piece of n-type semiconductor with p-type semiconductor, the "magic" occurs at the junction where the two types of semiconductors come into contact. All the "action" occurs within a thousandth of a millimetre of the border where the two materials meet.

On one side of the junction (in the n-type silicon) are electrons that are free to wander. On the other side of the junction are "holes", spaces in the silicon crystal lattice which are missing electrons.

1 Some electrons diffuse from n-type semiconductor to combine with holes in p-type semiconductor.

Recall that electrons and holes will readily recombine, if given the opportunity. They have such an opportunity at a p-n junction.

Initially, electrons from the n-type silicon cross the border and neutralise holes in the p-type silicon. This process eliminates mobile electrons and holes from a thin layer within the n-type and p-type silicon immediately next to the junction. This leaves the fixed negative charges within the layer of p-type semiconductor, and fixed positive charges within the n-type silicon, creating an electric field that pushes back electrons and prevents any more crossing the junction from the n-type silicon. The electric field also pushes back "holes" and prevents any more crossing the junction from the p-type semiconductor.

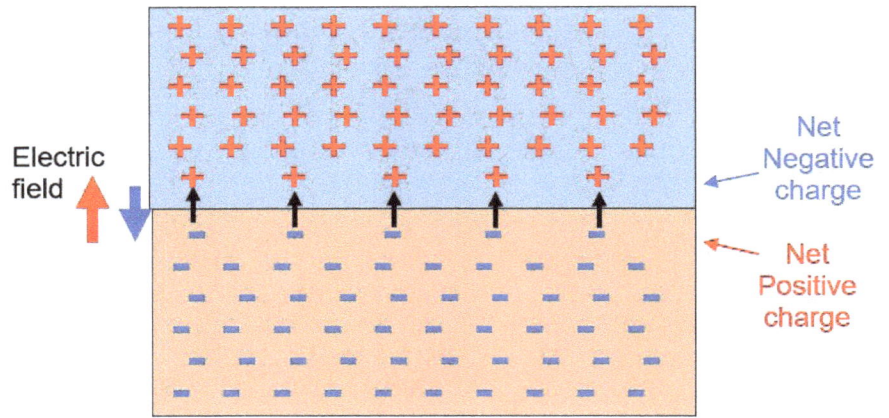

2   Transfer of electrons across the junction leaves a net negative charge in the p-type semiconductor, and a net positive charge in the n-type semiconductor. This creates an electric field pushing electrons back towards the n-type semiconductor, and holes back towards the p-type semiconductor. Combination of electrons with holes causes the region next to the junction to becomes depleted of both electrons and holes.

An equilibrium situation is reached where a thin region is formed around the junction which is devoid of electrons or holes. There are no charge carriers within this region, called the "depletion region". However, fixed negative charges within the crystal structure of the n-type part of the depletion zone, and fixed positive charges within the p-type part of the depletion zone, create a difference in electrical potential energy across the depletion zone of about 0.65 volts for silicon semiconductors (varying slightly depending upon the concentrations of phosphorus and boron dopants.

The depletion zone creates a "no man's land" for electrons and holes, preventing electrical current from passing through the zone. The 0.65 volt potential difference across the depletion zone is twenty five times as much as the average thermal energy of the electrons and holes. Only about one electron in a trillion has enough energy to cross the depletion zone. Accordingly, the concentration of free electrons at the "bottom of the depletion zone" (in the n-type semiconductor) is a trillion times larger than the concentration of electrons at the opposite side (in the p-type semiconductor). (Note 1)

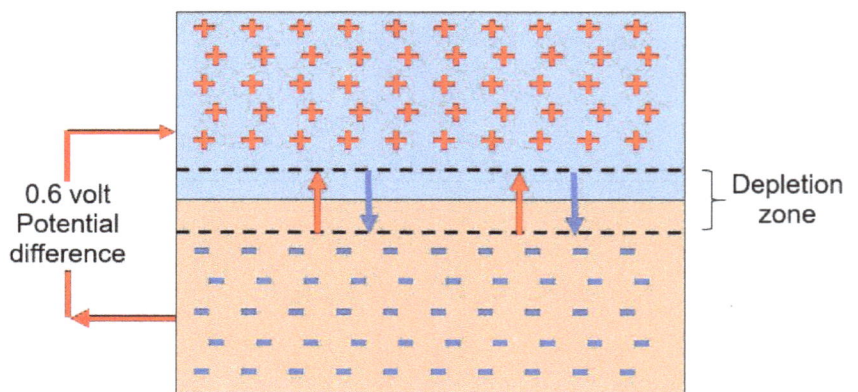

3   Electrons continue to cross the junction, causing the "depletion zone" to become wider. This causes the electric field pushing electrons and holes away from the junction to get stronger. Eventually, the energy required for electrons to pass through the electric field is as great as the energy liberated when electrons combine with holes (0.65 electron-volts). Then, no more electrons or holes pass through the junction. The region near the p-n junction is depleted of electrons and holes. Neither electrons nor holes have enough energy to push against the electric field to cross the p-n junction. The "depletion zone" has no mobile charge carriers which can conduct an electric current.

The 0.65 volt potential difference across the depletion zone represents a nearly impenetrable barrier – unless we reduce it by applying an external voltage to the p-n junction in the right direction.

Imagine that we connect the p-n junction to metal contacts, and connect the junction to a 0.4 volt battery, with the positive battery terminal connected to the p-type semiconductor. This pushes holes in the p-type semiconductor towards the junction, as well as pushing electrons in the n-type semiconductor towards the junction. The depletion zone is now much thinner than before. Furthermore, the potential difference across the junction has been reduced from 0.65 volts to 0.25 volts, and a measurable current starts to flow through the junction.

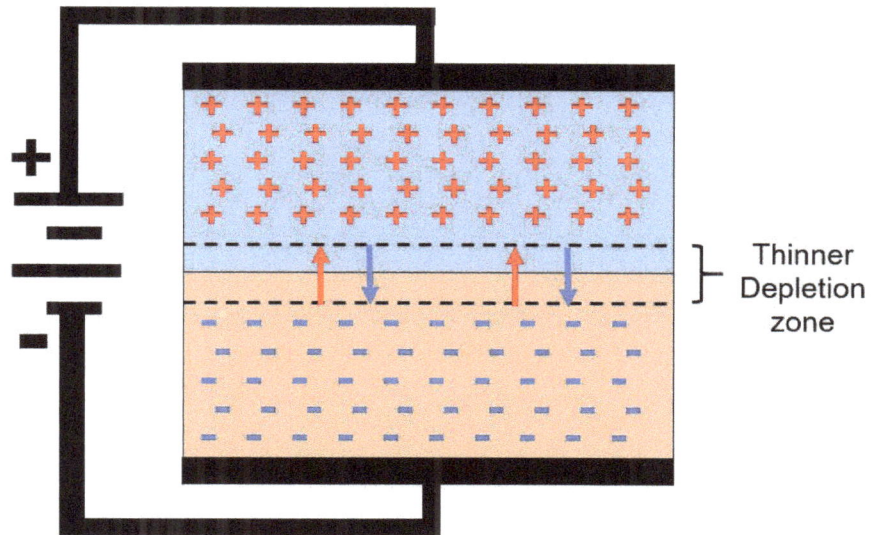

Thinner Depletion zone

If we connect the junction to a 0.7 volt battery, the depletion zone disappears entirely, and the junction region presents no barrier to current flow. Electrons have an unimpeded path through the p-n junction diode. Electrons flow from the negative terminal of the battery, into the n-type semiconductor, and cross the junction and combine with holes. On the other side of the p-type semiconductor, electrons transfer to the positively-charged electrode, forming new holes to replace those that are lost at the junction.

The diode is said to be "forward biased", and conducts as much electrical current as the battery and external circuit allows. In fact, we would need to introduce a resistor or other current-limiting device to prevent the p-n junction from overheating and burning out.

Now, let's imagine that we reverse the battery, so its positive terminal connects to the n-type semiconductor. The p-n junction is said to be "reverse biased". In this case, the applied battery voltage pulls electrons in the n-type semiconductor **away from the junction**, as well as pulling holes in the p-type semiconductor away from the junction. The depletion zone is now much thicker than before. Furthermore, the potential difference across the junction has been increased from 0.65 volts to 1.35 volts, so that **virtually no current** flows through the junction.[Note 2]

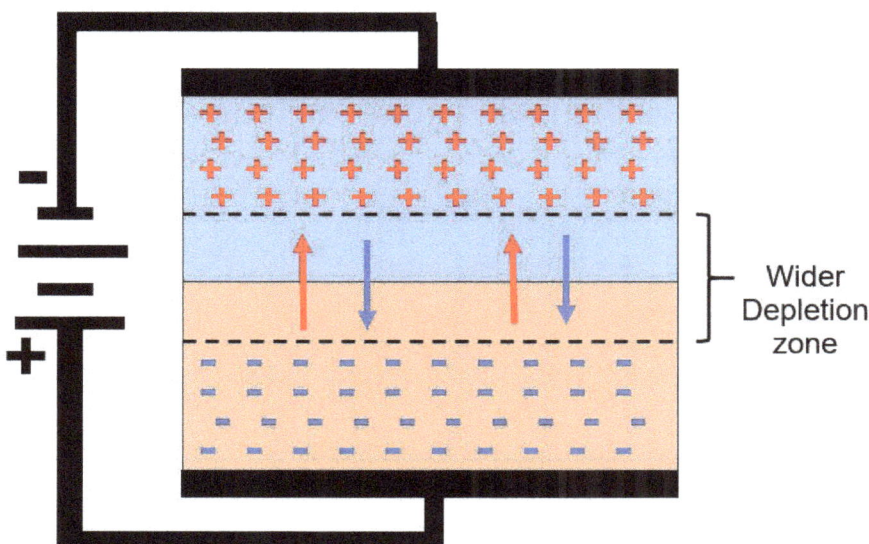

Wider Depletion zone

Thus, p-n junctions allow electrical current to flow unimpeded in the "forward" direction (provided that its 0.65 volt junction potential is exceeded), while blocking current flowing in the reverse direction. The first, and most obvious use, of p-n junction diodes was as a one-way valve, to convert alternating current, AC (alternating back-and-forth in direction) into direct current, DC. However, p-n junctions are now employed in many other critical applications.

Bear in mind that a p-n junction arises whenever a semiconductor is "doped" to create a surplus of electrons on side of the junction, and surplus of holes on the other side. Both the thickness of the depletion zone, and the potential difference across the depletion zone, depend on the level of doping of the n-type and p-type semiconductor.

The potential difference of about 0.65 volt across a silicon p-n junction arises from the amounts of phosphorus and boron that are added to make n-type and p-type semiconductor. As it turns out, it can readily be shown that the voltage difference ΔV across the depletion zone of a p-n junction is given by:

$$\Delta V = kT \ln\left[\frac{N_e N_h}{Ne_o^2}\right]$$

Where **k** is the Boltzmann constant
 T is the absolute (Kelvin) temperature
 kT has a value of 0.026 volts at 25°C,

 N$_e$ is the number of free electrons in the n-type semiconductor per cubic centimetre (equal to the concentration of added phosphorus atoms).

 N$_h$ is the number of holes in the p-type semiconductor per cubic centimetre (equal to the concentration of added boron atoms)
 Ne$_o$ is the number of electrons per cubic centimetre in pure ("undoped") silicon. This is also the number of holes per cubic centimetre.

Typically, the concentrations of phosphorus and boron added to make n-type and p-type silicon are about 10^{15}-10^{16} atoms per cubic centimetre. And, as we have seen, the concentration of electrons and holes in pure silicon are each about 10^{10} per cubic centimetre. Substituting these values into the equation above gives a voltage difference of 0.65 volt. If lower levels of "dopants" were used, the potential difference across a p-n junction would be less, and - as you might expect - if no doping is used, there would be no depletion zone and no potential difference.

The thickness of the depletion zone of a p-n junction also depends on the concentration of dopants added to make the n-type and p-type silicon. The greater the concentration of phosphorus and boron, the thinner is the depletion zone. With phosphorus and boron concentrations of 10^{15} atoms per cubic centimetre, the depletion zone is about one-thousandth of a millimetre thick. [Note 3]

Diodes and Light-emitting diodes

We have seen that a p-n junction allows an electrical current to pass in one direction only – when the p-n junction is "forward biased". Even then, a silicon-based diode conducts electricity only when an applied voltage is greater than about 0.65 volts. It takes energy to push electrons through the diode. This energy is normally converted to heat. The power (in watts) that is dissipated as heat is equal to the voltage difference (0.65 volts) multiplied by the electrical current (in amperes, or coulombs per second) passing through the diode.

Let's have a closer look at what happens when an electrical current flows through a p-n junction:

Current-limiting resistor

+
−

1. First, electrons flow from the negative terminal of a battery into the n-type semiconductor. Once there, these electrons are free to wander throughout the n-type semiconductor.

2. At the same time, electrons flow from the p-type semiconductor and out the opposite (positive) terminal, leaving additional holes behind. The formation of electron-hole pairs requires energy, and this energy is provided by the 0.65 volt potential difference across the diode. Once additional holes are formed, they can wander through the p-type semiconductor.

3. To prevent a build-up of charge, *electrons must pass through the p-n junction and recombine with holes in the p-type semiconductor. This recombination of electrons with holes releases energy*.

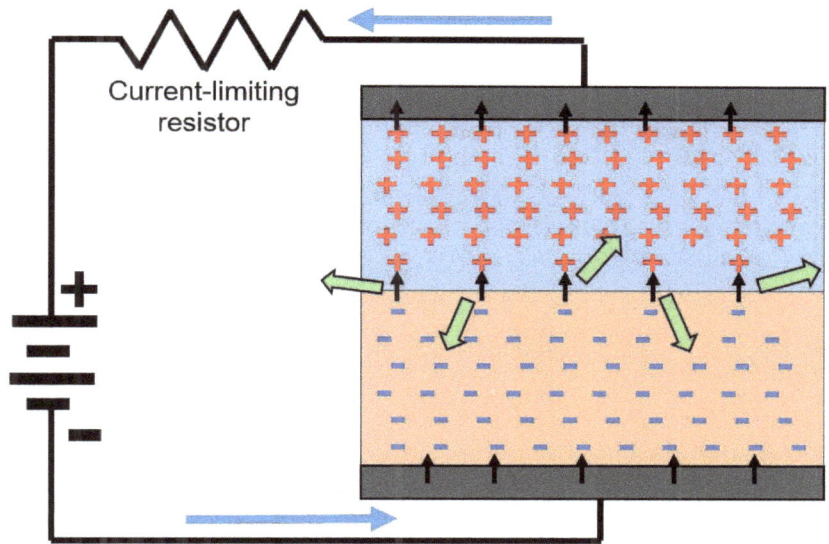

Energy released by recombination of electrons and holes at the p-n junction is converted into either light or heat. Normally, light that is emitted at the p-n junction of a diode is simply absorbed by the surrounding plastic case, and converted into heat. However, it is possible to design the p-n junction to optimise the production of light, and to allow light produced at the p-n junction to escape to the outside world. This is the concept of Light Emitting Diodes (LEDs), which have recently become widely used for household and commercial lighting. Eventually, LED lights might completely displace all other types of lamps (fluorescent tubes, halogen lamps, metal halide lamps, neon lighting).

As it turns out, silicon-based diodes are not suitable for producing Light Emitting Diodes. Energy released by electron-hole recombination in silicon is not sufficient to produce light at visible wavelengths. Different types of semiconductor material are used to produce red, yellow, green and blue LEDs. Red light is at the low-frequency end of the visible spectrum, and photons of red light have an energy corresponding to about 2 electron-volts. Blue light is at the high-frequency end of the visible spectrum, with a photon energy of about 3 electron-volts. Consequently, the operating voltage of LEDs typically varies between about 2 and 3 volts, depending upon its colour.

One of the first applications of LED lighting was for traffic lights. Throughout the 20th century, incandescent globes were extensively used for red, amber and green traffic lights. Incandescent lamps are very inefficient. Firstly, only a small fraction of their electrical power input is converted into visible light. And, of this, only about a third is within the desired red, amber or green colour range. Each lamp needed to be mounted behind a coloured filter, which absorbed light of all other colours. By contrast, LEDs are very efficient in converting electrical energy into light, and produce light only within the specified colour range. Consequently, LED traffic lights require only about 15% of the electrical power of conventional incandescent traffic lights. Energy savings are very substantial since traffic lights operate 24 hours/day, 365 days per year. Each intersection must have traffic lights pointing in each direction, and busy intersections can have several traffic signals facing in each direction.

Small indicator red and green LEDs have been available since the 1960s. Although the technology was gradually improved, the big breakthrough occurred in the mid-1990s, when the first high-brightness blue LEDs became available. This made possible, for the first time, manufacture of LED lamps producing white light. Two types of white LED light sources were developed. One incorporates red, green and blue LEDs together, producing light of the three primary colours which are combined to make white light. These are called "RGB" LEDs (for **R**ed, **G**reen, **B**lue). The second type uses blue LEDs to produce blue light which is absorbed by a phosphorescent compound, which then emits light photons of lower energy (in the red, yellow and green portion of the visible spectrum). This is very similar to how fluorescent lamps operate.

Solar modules

The same principle that underlies the operation of Light Emitting Diodes operates in reverse in solar photovoltaic modules. In this case, p-n junctions of very large surface area are illuminated with sunlight, and the light energy is converted into electrical power output. The electrical terminal on top of the p-n junction consists of thin metal contact wires, allowing sunlight to pass between the wires. Most solar modules are currently made from silicon.

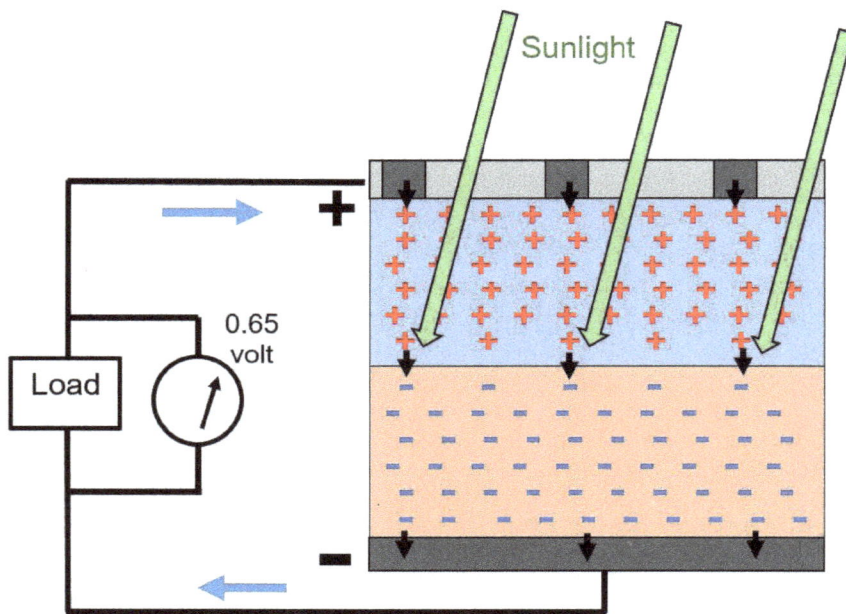

Operation of a silicon solar cell is exactly the reverse of a Light Emitting Diode. But here, solar light photons provide the energy to eject electrons from silicon atoms within the p-n junction. The resulting hole wanders into the p-type silicon. At the same time, electrons are pulled from the positive electrical terminal into the p-type semiconductor. This process is driven by the energy (0.65 volts) released by electron-hole recombination at the positive electrode. To prevent build-up of electrical charge, electrons flow out the negative terminal at the same rate as electrons are ejected from silicon atoms. Each silicon solar cell produces about 0.65 volts, and many cells are connected in series to produce solar panels producing 12 or 24 volts.

The voltage output of a solar panel remains relatively constant, so long as it receives significant levels of sunlight, however, the output current varies directly with the intensity of incident light. Thus, the power output of a solar panel varies directly with the intensity of sunlight.

The present generation of solar modules convert solar energy into electricity at about 20% efficiency. Some losses are due to reflection of sunlight off the front surface of the solar cell and terminal wire connections. However, the main loss is due to a fundamental limit on the amount of solar energy that can be utilised by a single-junction solar cell. For light to be converted into electricity, its photons must have sufficient energy to eject electrons from silicon at the p-n junction. Lower-energy photons (mainly in the infrared) are not absorbed, and cannot be utilised. Only light whose frequency (and energy) is above a certain threshold is absorbed, but photons with greater than threshold energy still produce the same 0.65 volts. Thus, only a portion of the energy of these light photons can be utilised.

This efficiency limitation can be overcome in multi-junction solar cells, typically incorporating three layered p-n junctions on top of one another. Each layer is made of a different semiconductor material. The top layer only absorbs light photons at the blue end of the spectrum, and thus, extracts energy from the most energetic photons in sunlight. The second layer absorbs light within the middle of the visible range, producing power from intermediate-energy photons. Light reaching the bottom third layer contains only lower energy photons (red or near infrared). In theory, a three-junction solar cell can convert up to 50% of incident solar energy into electrical power output, compared to a maximum theoretical efficiency of 30% for a single-junction silicon solar cell. Such multijunction or "tandem" solar cells are extremely expensive to produce, but are used in applications like space satellites and solar car racing, where high efficiency is extremely desirable.

Notes

1. The fraction of electrons (or holes) that have sufficient energy to cross the depletion zone varies exponentially with the potential difference **V** across the depletion zone as follows:

   Fraction of electrons with enough energy = $e^{-Vq/kT}$
   to cross depletion zone

   Where
   V is the potential difference across the depletion zone. This is 0.65 volts for a silicon p-n junction with no externally applied voltage.
   q is the charge on an electron, or a hole (1.6 X 10^{-19} coulombs)
   k is the Boltzmann constant
   T is the absolute (Kelvin) temperature
   At 25°C, **kT** is 0.026 volts

2. Even for a "reversed bias" diode, a tiny current does flow through the p-n junction. Don't forget that an n-type semiconductor does have some holes (at about one-millionth the concentration of free electrons), and a p-type semiconductor does have some electrons (at about a millionth the concentration of holes). When these "minority carriers" wander to the depletion zone, they are pulled to the opposite side of the p-n junction (by the same electric forces that push the "majority carrier" electrons and holes away from the junction).

3. For n-type silicon with concentration of phosphorus **Ne**, and p-type silicon with the same concentration of boron, the thickness of the depletion region can be calculated to be:

   $$\text{Thickness of depletion region} = 4\left[\frac{\varepsilon_0 \Delta V}{qN_e}\right]^{1/2}$$

   Where
   ε_o is electrical permittivity, which relates the force produced by a given electrical charge (8.85 X 10^{-12})
   ΔV is the electrical potential across the depletion zone (the potential difference produced by the p-n junction itself plus any external voltage applied across the junction).
   q is the electric charge of an electron (1.6 X 10^{-19} coulomb)
   Ne is the concentration of added phosphorus and boron (atoms per cubic metre).

20. Transistors

The electronic and computer revolution that transformed the world during our lifetimes arose largely from the invention of the transistor after the Second World War. These devices replaced the electronic "valves" (or, as Americans called them, "tubes") that had been used in radios, televisions, phonographs, etc. Like electronic "valves" before them, transistors are amplifiers. They amplify a very weak electrical signal (like that produced by a microphone or radio antenna) by thousands of times, producing a powerful output to drive, say, a speaker.

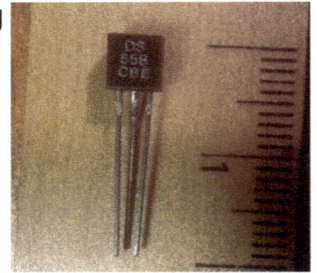

A transistor

Even the first transistors were very small devices, just a few millimetres in size, and dozens could be mounted on a printed circuit board. They were also relatively cheap, so it became feasible and economical to construct "logic circuits" to control all sorts of devices.

- There were timer circuits, giving a pause (varying from a fraction of a second to hours) between, say, pressing a button and something happening.

- There were circuits that would sense darkness and turn on the headlights of your cars;

- Circuits could automatically dim the headlights of a car when it sensed the headlights of another car approaching in the opposite direction;

- Circuits that would flash lights on your Christmas tree, or (now) your grandson's running shoes;

- Circuits that would control the speed of windscreen wipers, and even operate the wipers intermittently in a light rain.

Such circuits became more and more widely used, and more sophisticated. Rather than assemble such circuits on printed circuit boards, manufacturers began producing complex circuits within a single "chip", made from semiconductor silicon encapsulated within a plastic case. A single chip could contain a complex circuit containing hundreds of transistor amplifiers. As the technology improved, the number of transistors that could be contained within a single chip grew exponentially, doubling about every two years.

As electronic circuits within a single chip became increasingly complex, transistors undertook a new role. No longer were transistors mainly used to amplify weak signals. Transistors were increasingly being used as simple on-off switches in "logic circuits". The role of each transistor was trivially simple: When you receive a voltage signal (from the previous transistor), send – or don't sent – a voltage signal to the next transistor.

For example:

- Two transistors could operate together in an INVERTER circuit: It sends a voltage signal to the next transistor only if it receives *no* input voltage signal.

- A few transistors could operate together as an AND circuit: It sends a voltage signal to the next transistor only if the circuit receives an input voltage from *all* input stages.

- A few transistors could operate together as an OR circuit: The circuit sends a voltage signal to the next transistor if it receives an input signal from *any* of of its input stages.

It might seem to you that such logic circuits are incredibly simple and dumb. They do trivial tasks that the most stupid human could do easily. However, while individual logic tasks are extremely simple, each logic circuit can do millions of these tasks every second. Furthermore, putting together thousands - or millions – of these logic circuits on a single chip gives the capability to do all sorts of amazingly complex tasks.

So, let's look at the basic component which underlies all modern electronic and computer devices – the transistor. Remember that while its original role was as an amplifier, this role has been largely overtaken by its use in logic circuits. The transistor is now used mainly as a switch: it simply turns or, or turns off, when it receives an input signal. That's it. That's all it needs to do.

However, to do this job well, transistors in logic circuits must meet three critical requirements. Firstly, when a transistor receives a voltage signal produced by the previous the circuit, it should take hardly any electrical current or power from the previous circuit. The transfer of voltage signals ("hey you, turn on" or "hey you, turn off") should go only one way. Transistor A can send signals to Transistors B, C and D, but Transistor A should not be affected in any way by what Transistors B, C and D do.

Secondly, each transistor in a logic circuit must consume an absolute minimum amount of power. This is not simply about extending battery life, or being environmentally responsible. In a conventional application, like a radio, the transistor driving the loudspeaker might produce one or two watts of heat. That's no problem. But, if you have a *million* transistors on a chip, and each chip produces one watt of heat, that's a megawatt of heat. The chip would be instantly destroyed. Minimising power consumption is absolutely essential.

Fortunately, a transistor in a logic circuit doesn't have to do very much. It must simply send a voltage signal to turn on, or turn off, transistor(s) in the next circuit. So, how can a transistor switch on, or switch off, an output signal without consuming power? The answer is: Very Fast and Very Emphatically. The transistor is either switched fully off (like an open circuit) or switched fully on (like a short-circuit).

- In the "off" state, as an "open circuit", the transistor consumes no power because no current passes through it.

- In the "on" state, as a "short circuit", the transistor consumes no power because it doesn't impede the flow of current (so there is no voltage across the transistor).

Logic circuits are designed for the transistors to either be fully OFF, or fully ON. Transistors consume power only when they are partially on, during the billionth of a second or so required to switch between "on" and "off". So, the third requirement for transistors in logic circuits is that they must be able to switch on, or switch off, very fast.

Modern transistors meet these requirements pretty well. When switched ON, they provide a very low resistance path that is virtually a short circuit. When switched OFF, they provide a very high resistance path that is virtually an open circuit. Their switching times are very short. Even so, the gazillions of transistors in servers and data centres located around the world (which comprise "the cloud"), switching billions of times per second, use an estimated 5% of all electrical power produced in the world

The Big Switch

So, how do transistors switch electrical currents on and off?

The first transistors were "bipolar" transistors. Bipolar transistors are still widely used, but they have now been largely displaced in logic circuits by "Field Effect Transistors" (FETs). Although the technology required to produce FETs is more complex, and their introduction lagged "bipolar transistors" by at least a decade, the basic operating principle of FETs is easier to understand.

The first FETs were "Junction Field Effect Transistors" (JFETs). These rely on the same p-n junctions that are utilised in diodes.

Consider a long, thin slab of n-type silicon, with a layer of p-type semiconductor on opposing sides. As you would expect, the slab of n-type silicon contains electrons that are free to migrate and conduct electricity through the silicon crystal (which contains equal numbers of positive charges that are fixed within the crystal structure). The slab of n-type semiconductor provides a channel for electric current to flow through it. Electrical contacts at each end of the channel (termed the "Source" and "Drain" electrodes) allow current to flow into and out of the channel from an external electrical circuit.

Each p-type layer also has an electrical contact (termed the "Gate" electrode). The Gate electrodes on both side of the channel are electrically connected to each other (and so, are at the same electrical potential).

At each junction between n and p-type silicon, a depletion zone arises. The depletion zone does not necessarily extend to the same depth within the n-type and p-type semiconductors. Since the n-type semiconductor is made with a much lower concentration of dopant than the p-type layer, the depletion zone extends a much greater distance within the n-type channel than in the p-type layer.

Because the n-type slab is very thin, and because it is so lightly "doped", the depletion zone normally extends across the n-type channel. In this state, there are virtually no free electrons within the channel to conduct electricity. Consequently, the "n-channel" JFET is effectively an open circuit, providing a non-conducting path

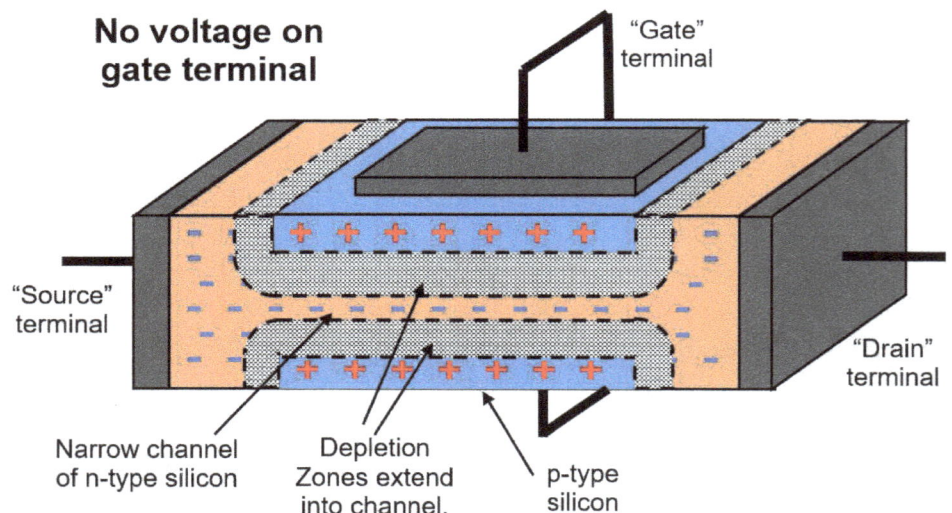

between the Source and Drain electrodes, *when the gate electrode is at the same potential as the Source*.

We can change this situation by applying a small electrical voltage across the p-n junctions through the Gate electrode(s). By applying a voltage that makes the Gate positive relative to the Source - we reduce the thickness of the depletion zone, exactly as occurs when we "forward bias" a p-n junction in a diode. This "opens up" the n-type channel, which is then able to conduct electricity. The channel then provides a low resistance path for electrical current to flow from the Source to the Drain (exactly analogous to water flowing from a tap into a sink, and then down the drain pipe).

Thus, the "n-channel" JFET is essentially a short-circuit, providing a conducting path between the Source and Drain, *when the Gate is at higher voltage (positive) relative to the Source*.

JFETs are also made using p-type semiconductor for the channel

Gate terminal at higher voltage than Source

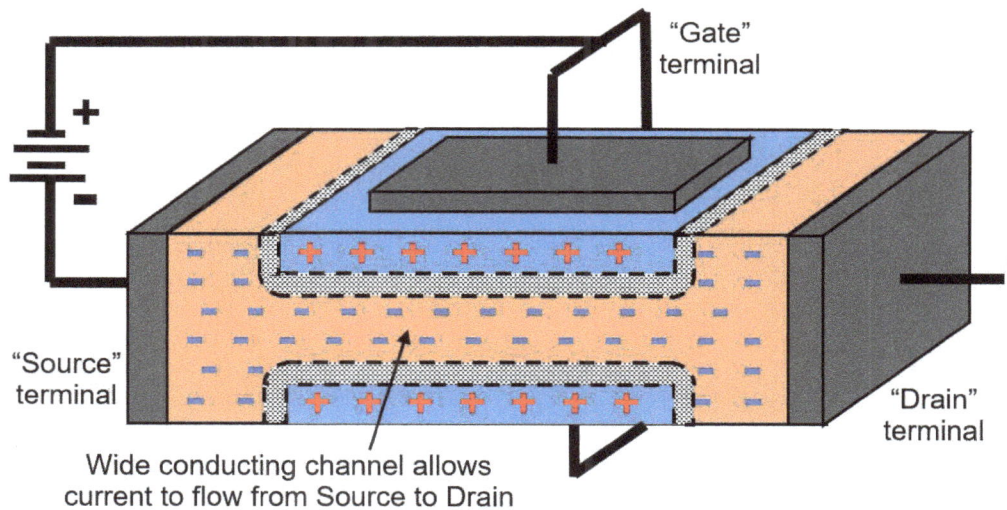

"Gate" terminal

"Source" terminal

"Drain" terminal

Wide conducting channel allows current to flow from Source to Drain

between Source and Drain terminals, with layers of n-type semiconductor on each side. This makes a "p-channel" JFET. These operate in exactly the same way as "n-channel" JFETs – except that *"p-channel" FETs are "turned on" when the Gate terminal is at lower voltage (negative) relative to the source.*

More recently, Junction FETs have been largely replaced by "Metal Oxide Semiconductor" FETs (MOSFETs). Their basic operating principle is even easier to understand. Like their JFET cousins, MOSFETs employ a long, thin slab or n-type or p-type silicon as a conducting channel between a Source and Drain electrode.

Let's consider a MOSFET containing a channel of n-type silicon. The conductivity of the channel is controlled by a Gate electrode, which creates a depletion zone within the adjacent channel. Instead of using a p-n junction to create the depletion zone, MOSFETs use electrostatic repulsion or attraction of electrons across a very thin insulating layer. The Gate electrode is separated from the channel by a layer of silicon dioxide, which is an excellent insulating material (essentially the same as glass).

n-channel MOSFET

Channel is depleted of electrons when Gate is negative, or at same potential, as Source.

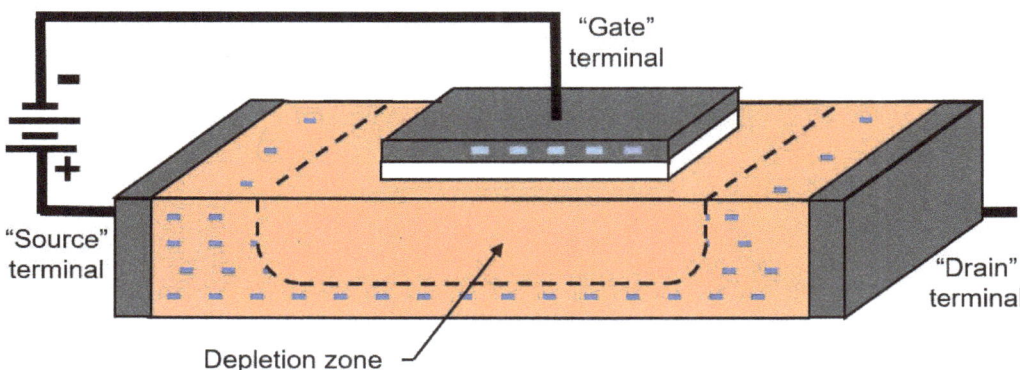

"Gate" terminal

"Source" terminal

"Drain" terminal

Depletion zone

When the Gate is at lower voltage (negative) relative to the Source, electrons are repelled from the n-type channel away from the Gate, causing a depletion zone to extend across the channel. Without free electrons within the channel, it becomes non-conducting.

Most n-channel MOSFETS are designed so that the channel is non-conducting even when the Gate is at the same voltage as the Source.

In order for an n-channel MOSFETs to turn ON, the GATE electrode must be at higher voltage (positive) relative to the Source. This attracts electrons into the channel alongside the Gate. With free electrons present, the channel can conduct an electric current between the Source and Drain.

n-channel MOSFET

Wide channel conducts current when Gate is positive relative to Source.

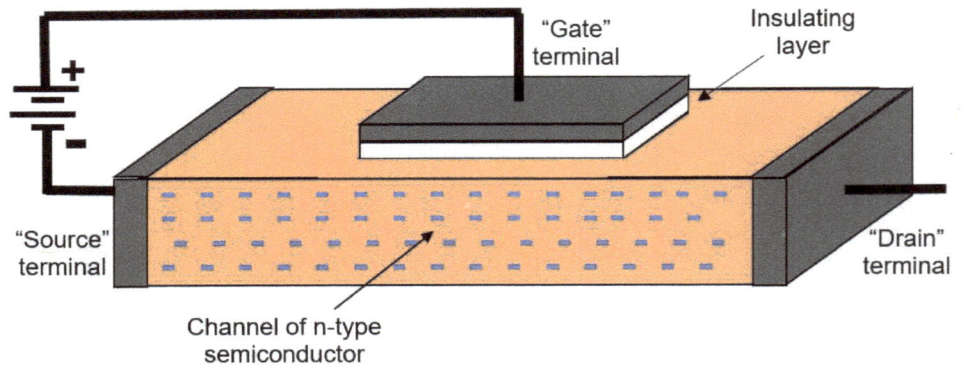

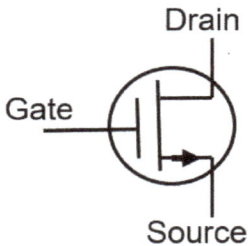

n-channel MOSFET
- Turned off when gate is at same potential as source.
- Turned on when gate potential is positive (+) relative to source.

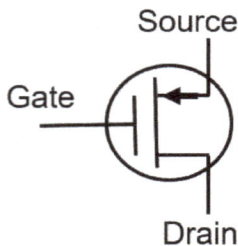

p-channel MOSFET
- Turned off when gate is at same potential as source.
- Turned on when gate potential is negative (-) relative to source.

The symbols used to represent MOSFETs indicate how they work. The Gate electrode is shown parallel to the Source-Drain channel, with a gap representing the thin insulating layer between the Gate and the channel. An arrow is used to indicate the direction of current flow in n-channel and p-channel MOSFETs. The symbols and basic characteristics of n-channel and p-channel MOSFETs are shown here:

The characteristics of n-channel and p-channel MOSFETs are opposite, and they complement each other. In fact, many integrated circuit chips contain both n-channel and p-channel MOSFETs in series. Application of an input voltage to the gates of both MOSFETs ensures that one is switched fully ON, and other is switched fully OFF.

Consider for example, the operation of n-channel and p-channel MOSFETs in a simple "inverter" circuit.
- When a positive input voltage is applied to the gates of both transistors, the p-channel MOSFET is turned fully OFF. The n-channel MOSFET is turned fully ON, providing an effective short-circuit connecting the output of the inverter to ground (zero voltage).

- When zero voltage (ground potential) is applied to the gates of both transistors, the n-channel MOSFET is turned fully OFF, disconnecting the output from ground. The p-channel MOSFET is turned fully ON, connecting the output of the inverter to the positive output voltage (in this case, +10 volts).

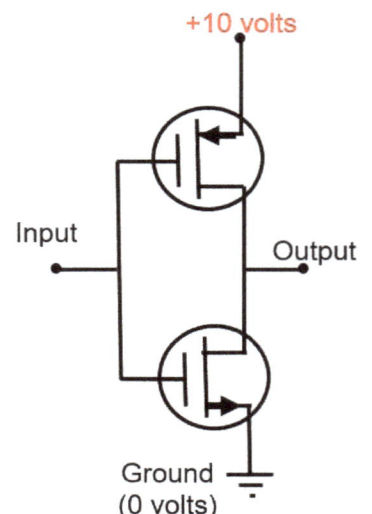

104

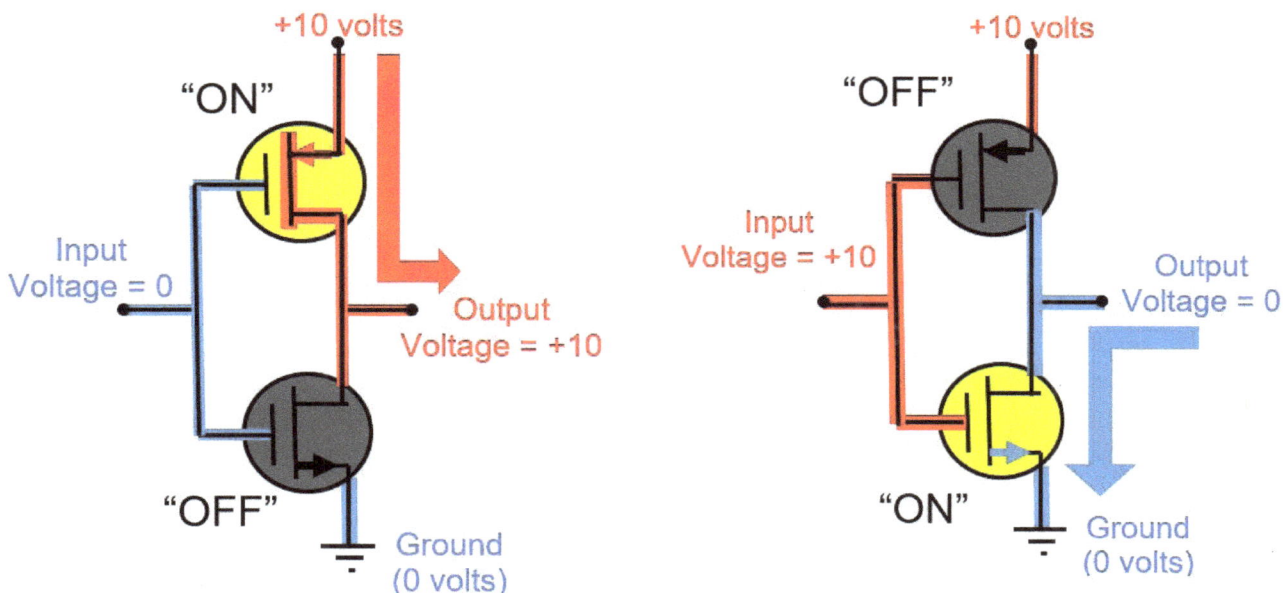

So, when the input to the inverter is "low" (at ground potential, or zero volts), its output is "high" (at +10 volts). When the input to the inverter is "high", its output is "low". The output voltage is always the inverse of the input.

Many integrated circuits are now based on n-type and p-type MOSFETs in complementary circuits. These types of logic circuits are called "CMOS", or I have heard them referred to as "COSMOS". This stands for "**CO**mplementary **S**ymmetry **M**etal **O**xide **S**emiconductor".

A MOSFET integrated circuit chip

A simple, brief overview of semiconductors, MOSFETs and CMOS is given by the following 7 minute video:
https://www.youtube.com/watch?v=8Np0qawbbCI

Here is an 8-minute video reviewing MOSFET operation and explaining how they are used in logic circuits.
https://www.youtube.com/watch?v=J8ZPIDNaijs

Modern computers do all calculating and processing data in "digital", or "binary", form. All input data is in "bits" – a signal that is either "On" or "Off". An "On" signal generally consists of a pulse at the positive supply voltage (say, +10 volts), and an "Off signal" comprises zero voltage (ie. at "ground potential"). Signals are usually represented as a "1" (at the positive supply voltage) or a "0" (at zero voltage).

We humans are familiar with a ten-based numbering system, and this is how we like to input data into a computer. Computers convert these numbers into digital form (a series of "bits"). There are two possibilities for one bit (either a 1 or 0), four possibilities for two bits (either 1-1, 1-0, 0-1 or 0-0). There are 256 possibilities for eight bits, and this is more than enough for each particular combination to uniquely represent a single letter, digit and symbol on a keyboard. Eight bits is called a "byte", and this is the unit used to specify the capacities of memory devices – usually in megabytes (a million bytes) or gigabytes(a billion bytes).

21. Buildings and structures: Icons of modern society

Introduction

Growing up on New York City, I was surrounded by iconic buildings and structures. Where my family lived in lower Manhattan, we could (and did) walk to the Williamsburg Bridge on Delancey Street, and would sometimes walk over the bridge to Brooklyn. Not far away were the Brooklyn and Manhattan Bridges. All were built in the decades leading up to, and just after, the beginning of the 20th century, and these bridges were instrumental in allowing the city to extend from Manhattan into the "bedroom boroughs" of Brooklyn and Queens.

Starting around 1880, many cities of the world were transformed by the construction of buildings and structures that were far bigger and technically challenging than anything built before. This was driven largely by development of the blast furnace and Bessemer Converter which, for the first time, enabled steel to be produced in large quantities at relatively low cost.

Of course, iron and steel had been produced for millennia, but high cost and limited quantities restricted their use mostly to tools and weapons. Once steel became available in large quantities with consistent quality, this created new opportunities for construction. Steel offered much greater tensile strength than masonry materials or timber, enabling new types of structures and larger structures to be built that were simply not feasible to build before. Imaginative engineers seized the opportunity, and captured the imagination of the world. The Eiffel Tower was built to showcase France's industrial prowess at the Paris Exhibition of 1889. The Eiffel Tower was intended to be dismantled at the conclusion of the Exhibition, but was embraced by the people of Paris and millions of visitors from around the world. It still is.

When it was built, the Eiffel Tower was the tallest man-made structure in the world. It remained the tallest structure until that record was seized by construction of the Chrysler Building in New York some forty years later.

Most of the time that my family lived in lower Manhattan, we didn't own a car. On weekends, we often went for long, all-day walks. Sometimes we would go through Chinatown, Little Italy, Greenwich Village, Central Park, the Museum of Natural History, or take the ferry from the tip of lower Manhattan to Staten Island. After walking all day, we would often arrive unannounced at the apartment of my father's brother, where my aunt would cater an impromptu dinner, much to my mother's embarrassment and annoyance.

One of our favourite landmarks and places to visit was the Empire State Building, where my father worked for most of his life, delivering parcels to hundreds of offices and small businesses within this one building.

The Empire State Building was built around 1930, at the same time as the Chrysler Building. This was a period when building "Skyscrapers" was a great source of pride. The Empire State Building and Chrysler Building were locked into a competition to be the tallest building on Earth. Both builders kept their plans, and the intended height, a closely-guarded secret. The Empire State Building won the competition. It was the tallest building in the world, and held that title for a generation. Since then, numerous taller buildings have been constructed, but none remained the tallest building very long or captured the imagination of the public in the same way.

New York, like most major cities of the world, was first setteld because its river and harbour provided a lifeblood for transport and trade with other major centres around the world. However, as the city grew over the following decades and centuries, the rivers, harbours and inlets that were so essential to its existence began to constrain its growth. As the city sprawled across the East River into Brooklyn and Queens, ferries carried the growing traffic of people and goods across the river, but it must have been very clear that ferries were not a long-term practical solution. A better method was needed to allow large numbers of vehicles and people to cross the river quickly, easily and cheaply. The construction of the Brooklyn Bridge in the latter half of the nineteenth century was a "game-changer". It must have been a great sensation of the age. In the following decades, many other suspension bridges were built across the East River, Harlem River, Hudson River and other waterways around New York. Even as I was growing up in lower Manhattan, in the mid-20th century, the three bridges within walking distance of our home (the Brooklyn, Manhattan and Williamsburg Bridges) were iconic features of the skyline and a palpable sense of pride.

In the 1960s, shortly after my family moved to Brooklyn, I watched the Verrazano Narrows Bridge being constructed from the lounge room window of our apartment on the 13th floor. At the time, the 1.3 kilometre central section of the bridge was the longest suspended span in the world. The Verrazano Bridge is still the longest suspension bridge in the United States. A few years after it was constructed, I drove over the bridge each day to commute to my first full-time job in New Jersey. One sunny day, as I drove up the long approach ramp to the bridge, I encountered the most amazing sight: the roadway completely disappeared into the clouds!

About a decade later, the World Trade Centre was built in lower Manhattan, and my brother and I were not impressed. The two towers had taken the title of tallest building in the world, but we thought that they were just two huge rectangular boxes. They had none of the elegance and art deco style of the Empire State Building. Our opinion changed during a visit to New York in 1996, when my wife, kids, brother and I happened to be walking past the World Trade Centre one late afternoon, just as dusk approached. We decided to visit the observation deck at the very top of one tower. Climbing a flight of stairs onto the open deck on the roof, I was literally stunned by one of the most amazing views I had ever seen. It was like a 360-degree view from the top of the world.

Five years later, I was awoken by a telephone call from my daughter in England, telling me to turn on the television. Like millions of others, I watched as the World Trade Centre towers burned and collapsed. I had no idea what consequences would follow, but I knew that I was watching an epoch-making event and that the world would never be the same.

22. Materials and structures prior to the industrial revolution "The age of compression"

Since the beginning of recorded history (and before) mankind has been building huge monuments, temples, tombs and fortresses. The predominant material for building such structures has been stone and cements (derived from minerals) until the 1800s, when technology became available to produce iron and steel relatively cheaply and in large pieces. Consequently, until recently in human history, the design of major buildings and structures was constrained by the limited strength of stone, cement and concrete. If we want to understand why and how such structures were designed, and how modern structures are designed, we need to understand the properties of the building materials available.

Stone has been quarried since prehistoric times, and cement and concrete were used in ancient Rome. Different types of stone, cement, mortars and concrete vary significantly in strength, but all have similar types of properties. Stone and cement have a reasonably good ability to resist compressive stress – force pushing inwards per unit area of the material.

As you might expect, the ability of a material to withstand crushing varies directly with the cross-sectional area to which the force is applied. Consequenlty, the compressive strength of a material is expressed as the maximum **force per unit area** that can be applied before the material crushes or collapses. Note that the compressive strength of a material (like its tensile strength and shear strength) has the same units as pressure, and is usually measured in units of megapascals (MPa, where one MPa is one million Newtons of force per square metre). I'll use the symbol **Pc** for compressive stress (compressive force per square metre) and Pc_o for the critical compressive stress at which the material crushes.

While the compressive strength of stone and concrete is not nearly as high as steel, masonry structures tend to have thick walls (over which compressive forces are spread), so walls and columns of stone are generally quite good at resisting the downwards force of large weights. On the other hand, stone and concrete have extremely poor ability to resist tensile forces pulling them apart. As a rough rule of thumb, concrete has about thirty times less compressive strength than steel, but its tensile strength is **three-hundred times** less than steel.

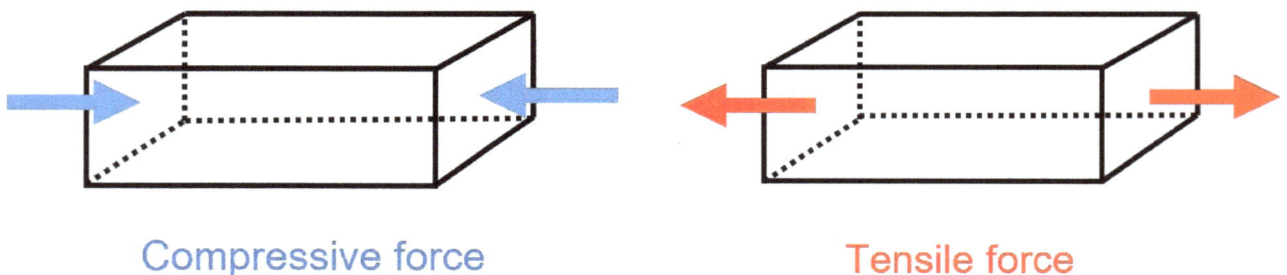

Compressive force   Tensile force

As a consequence, stone structures are hardly ever designed to withstand tensile forces: they are nearly always designed to support compressive load forces. Until the latter half of the nineteenth century, buildings, monuments, bridges and other major structures were designed so that their structural parts operated in compression. Thus, we can think of the period up to the industrial revolution as being the "Age of compression".

Imagine that you are the king, queen, or pharoah or ruler of a kingdom and decide to build a huge monument to the greatness of your reign (as rulers often did, and perhaps still do). Let's say that you decide to construct a cylindrical tower to the heavens constructed of concrete. Let's consider how high you could build such a column.

Let's say that the column has a cross-sectional Area **A** and height **H**. The total mass of the column is given by its density multiplied by its Volume (equal to its area **A** multiplied by its height **H**).

Concrete has a density **ρ** (about 2,300 kilograms per cubic metre), so the mass of the column is:

Mass of column = ρ A H

The weight of the column – that is, the downwards force of attraction towards the Earth - is equal to its mass times the acceleration of gravity **g**.

So, downards force of weight of column = ρ A H g

The downwards force of gravity is spread across area **A** at the bottom of the column (where the compressive force is greatest). So, the **compressive force per unit area** (the "compressive stress") is:

$$\text{Downwards force per unit area} = \frac{\rho \cancel{A} Hg}{\cancel{A}} = = \rho\, Hg$$

In order for the column to not be crushed by its own weight, the downwards force per unit area must be less than the compressive strength **Pc$_o$** of the concrete.

Condition for column to not collapse $\rho\, H\, g < Pc_o$

Let's solve this equation to find the maximum possible height for a cylindrical column of concrete. The density of concrete ρ is about 2,300 kilograms per cubic metre, the acceleration of gravity at the surface of the Earth **g** is about 10 metres/second^2, and the compressive strength of concrete **Pc$_o$** is about 30 million Pascals (30 million Newtons of force per square metre). Substituting these numbers into the equation gives a maximum height of about 1,300 metres. That's 1.3 kilometres! Even the most egocentric rulers in all of history did not built monuments this high. And, even this is not an absolute limit. We could build a concrete or stone structure even higher if it is pyramidal shape, rather than cylindrical, so that the weight of the structure is spread over a greater area at its base. Thus, we can safely conclude that the compressive strength of concrete or stone was not a limiting factor constraining the vertical height that concrete or stone walls could be constructed.

The problem faced by egocentric rulers in ancient times was not the height of walls and structures that they built, but the horizontal distance that could be spanned by a roof on the structure. If they constructed a flat horizontal roof, or had horizontal beams made of concrete or stone, this would give rise (as we shall see) to high tensile forces acting on the bottom surface of the beam. These tensile forces increase with a longer beam or a wider span of the roof. To prevent such stone beams from breaking, the ancient Greek architects would have to severely restrict the distance spanned by the beams. In designing their temples and other grand buildings (such as the Parthenon, for example), the Greeks used many closely-spaced columns, so that the distance spanned from one column to the next was very short.

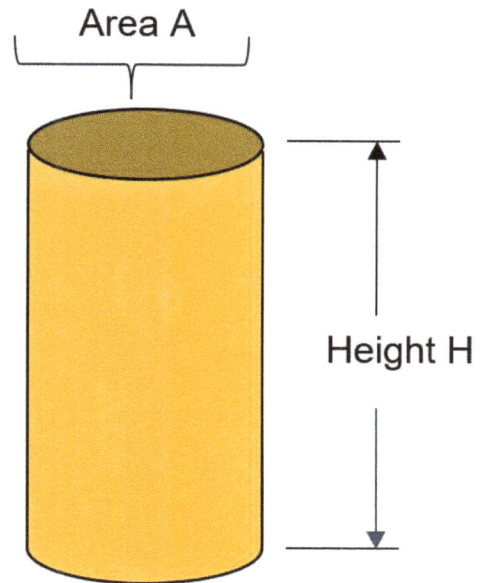

The Parthenon in Athens, Steve Swayne, httpscommons.wikimedia.orgwikiFileThe_Parthenon_in_Athens.jpg

Later, the Romans and others used a different design strategy to avoid structural failure of their roofs and bridges due to tensile forces. The approach was quite simple: structures were designed so that the concrete or stone in the roof or span was always subject to compressive

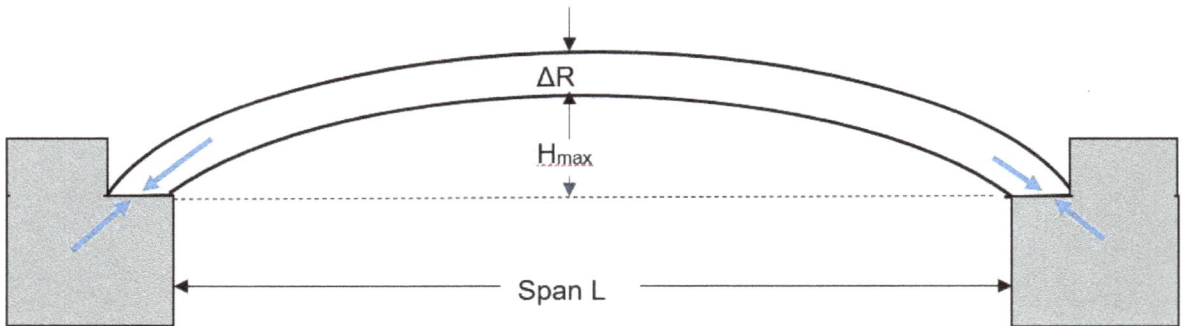

forces: and was never subject to tensile forces. This could be achieved by using an upwards arch for a bridge or the roof of a structure. "Romanesque architecture", which was common throughout the Middle Ages, was characterised by semicircular arches over windows and doorways. Many arches are relatively flat, spanning a distance **L** that is much longer than the maximum height mid-way along the arch H_{max}. The arch is a small segment of a huge circle, with a line connecting the two ends of the arch being a "chord" of the circle. This situation is perhaps the easiest to understand and analyse, and is the type that I will consider here.

Let's consider the forces acting on such an arch, where the distance **L** spanned by the arch is much less than the radius of the arch.

Each half of the arch can be viewed as a "slice" of the circle, subtending an angle **θ** at its centre.

Imagine that the arch is made of concrete or stone with a density ρ, which typically has a value of about 2,300 kilograms per cubic metre. The length along the surface of the arch depends on the radius **R** and angle θ and, in fact, is equal to the product **Rθ** (where angle θ is expressed in radians).

We can consider each half of the arch to be a rectangular slab of material which has been bent to follow a circular contour. Its volume is simply the length (**Rθ**) multiplied by its width (**W**) multiplied by its thickness **ΔR**. To get the mass of the arch, we multiply its volume by the density ρ of the material.

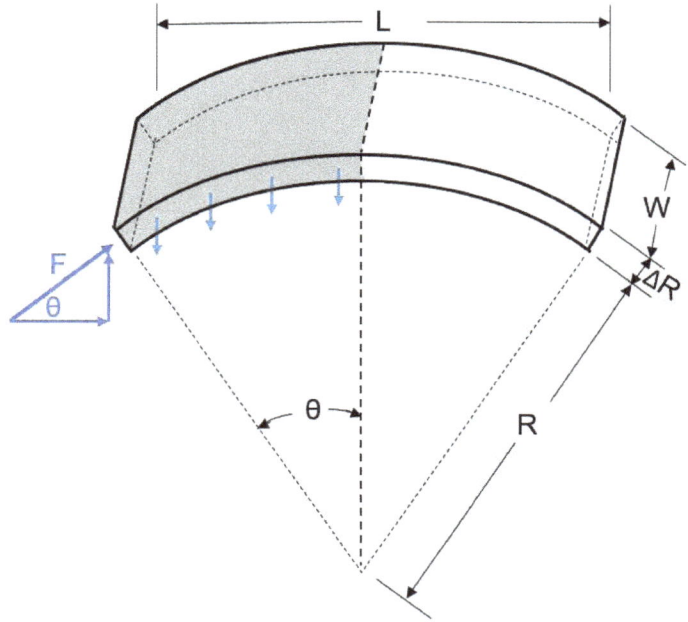

Mass of arch = ρ [(Rθ) W ΔR]

Finally, to get the **weight** of the left-hand-side of the arch (the force of gravity pulling the arch downwards), we multiply its mass by **g**, the acceleration of gravity at the surface of the Earth (9.8 metres/second²).

Weight of arch = ρ [(Rθ) W ΔR] g

Now, the only thing holding up the weight of the slab is the inwards compressive force **F** pressing on the ends of the arch. The compressive force acts in the direction of the arch, that is, at angle θ from horizontal. It has a horizontal component (pressing inwards) and an upwards component equal to **F sinθ**.

If the arch is relatively flat, angle θ will be relatively small. In this case, we can use a very useful mathematical simplification. For small angles, **sinθ** is nearly equal to θ (where the angle is in units of radians). Consequently, we can substitute angle θ for **sinθ** (with an accuracy of 95% for angles up to 30 degrees).

Since the arch is not moving, all forces acting on the arch must be balanced. So, the upwards component of the compressive force acting on each side of the arch must be equal to the weight of that half of the arch. This means that:

$$ F\,\theta \; = \; \rho\,[\,(R\theta)\,W\,\Delta R\,]\,g $$

Upwards component of compressive force

Weight of arch

Note that angle θ appears on both sides of the equation and cancels out.

We can express the compressive force **F** in terms of the compressive strain **Pc** (the compressive force per unit area). The cross-sectional area of the arch is equal to **W ΔR**, so the compressive strain is given by:

Equation (1)   Compressive strain in arch, $Pc = \rho R g$

Interestingly, if the arch is relatively flat – as we have assumed – *the compressive stress has the same value at every point along the arch*.

We can build the arch as long or as flat as we like – provided that the compressive stress does not exceed the critical compressive strength Pc_o of the material comprising the arch. Otherwise, the material in the arch will crush and the arch will collapse.

The problem with Equation (1) is that, if we are building an arch, we wouldn't care about the radius of the arch per se. Rather, the important design criteria are the distance **L** spanned by the arch, and the maximum height of the arch at its centre H_{max}. Using simple geometry, we can easily derive an equation expressing the radius of an arch in terms of its span **L** and maximum height H_{max} midway along the arch. For the interested (or sceptical) reader, I derive this relationship at the end of this chapter [Note 1]. The result is:

$$\text{Equation (2)} \quad \text{Radius of curvature of arch} = \frac{1}{8}\frac{L^2}{H_{max}}$$

Substituting Equation (2) into Equation (1) tells us how the compressive stress along the arch varies with its span and maximum height:

$$\text{Equation (3)} \quad \text{Compressive stress } Pc = \frac{1}{8}\,\rho\,g\,\frac{L^2}{H_{max}}$$

If the arch is not to collapse, which would be very bad for the reputation of its designer (and worse for anyone standing on the arch), we must ensure that the compressive stress within the arch is less than the compressive strength of the material from which it is made. In other words, the designer must ensure that:

$$Pc_o < \frac{1}{8}\,\rho\,g\,\frac{L^2}{H_{max}} \quad \text{Where the symbol } < \text{ means "is less than"}$$

This sets a maximum span for the arch L_{max} in terms of its maximum height H_{max} mid-way along the arch and the compressive strength and density of the material from which it is made:

$$\text{Equation (4) Maximum possible span } L_{max} = \sqrt{\frac{8\,Pc_o\,H_{max}}{\rho\,g}}$$

Where Pc_o is the compressive strength of the material of which the arch is made
H_{max} is the maximum height of the arch at mid-span
ρ is the density of the material of which the arch is made
g is the acceleration of gravity at the surface of the Earth (9.8 metres/sec²)

To achieve the maxiumum possible span for an arch bridge, we should use a highly curved arch with a large maximum height at its centre, but this poses a range of problems. For one thing, the simplifying assumption used in our analysis of a "flat" arch become less applicable, and Equation (4) might no longer be accurate.But there are practical reasons to limit the maximum height of the arch. For example, if we were building a bridge, we would want to limit the slope of the bridge to an angle that would be suitable for cars, horse-drawn carriages or whatever.

Let's say that we decide to restrict the maximum height to one-fortieth the span, which would would give a slope angle of 6 degrees at the ends of the bridge. We substitute $H_{max} = 1/40 \ L_{max}$, 2,300 kg/m³ for the density of concrete ρ, and 30 X 10⁶ MPa for the compressive strength of concrete Pc_o. Under these conditions, the maximum span of a unreinforced concrete arch is 260 metres. Of course, this is an absolute maximum, and no sane engineer would ever design a bridge that is only just capable of supporting its own weight and, even then, is at the threshold of collapse.

Victoria bridge in Brisbane, with its central arch spanning the centre of the river.

So, in principle, arches can span considerable distances. However, to maintain the compressive force in the arch (which holds up its own weight), the arch must push outwards against the supports at both ends. An arch bridge spanning a river will push outwards against the structural anchors on the banks of the river (or, against adjoining arches, with arches on the end pushing against the structural anchors). An arch spanning the roof of a building will push outwards against the walls on which it rests. These sideways forces can be quite large, so if an arched roof covers a building, the walls must be braced to withstand the outwards sidways thrust exerted by the arch. If the walls or anchors of the arch cannot maintain the compressive force required to support the arch, the arch will collapse.

In Medieval times, cathedrals were often built with "flying buttresses" projecting outwards to brace the walls and prevent them from being pushed outwards by the compressive forces generated by arched or peaked roofs.

It is important to understand that the compressive forces within an arch are not just an annoyance and limitation in the design of bridges, buildings and other structures. ***Compressive forces along the arch generate the upward force that supports the weight of the arch***. This occurs not only at the ends of the arch, but ***at every section along the arc***. Let's consider how this occurs.

Consider, as before, a "flat" arch which is a thin section of a circle. The slope of the arch gets steeper and steeper as we move away from the centre of the arch. It is this continuous variation in slope over the length of the arch that generates an upwards force.

Let's imagine one section of the arch (shaded in green) with a small length ΔL and constant cross-sectional area **A**.

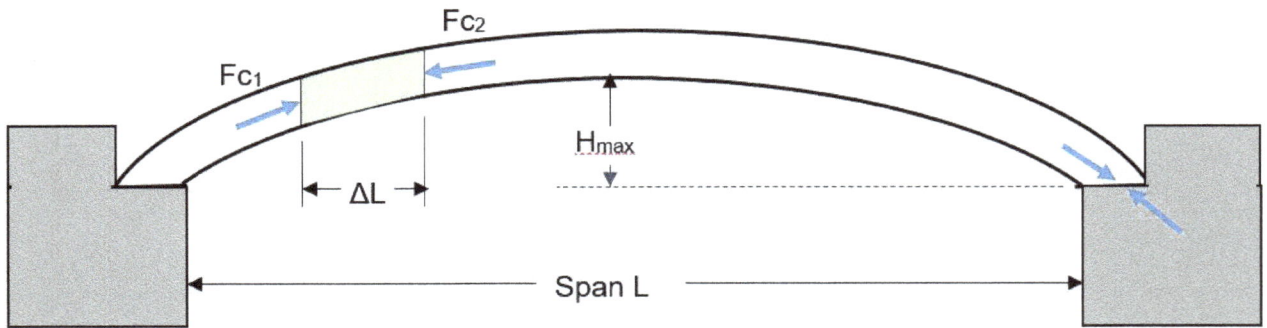

The weight of the arch gives rise to a compressive force along the length of the arch. Consider the compressive forces Fc_1 and Fc_2 acting respectively on the left and right faces of the section. Note that these forces differ slightly in their angle θ relative to the horizontal due to curvature of the arch

Let's look at an expanded view of compressive forces Fc_1 and Fc_2, and consider the horizontal and vertical components of these forces.

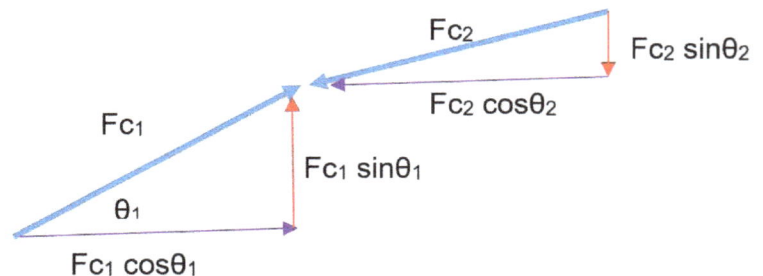

Compressive forces Fc_1 and Fc_2 oppose each other. Since there must be no net horizontal force on this (or any other) section of the arch, the horizontal component of these opposing forces are equal. For those who remember their triginometry:

$$Fc_1 \cos(\theta_1) = Fc_2 \cos(\theta_2)$$

> Where θ_1 is the angle of the arch from horizontal at the left-hand side of the section
> θ_2 the angle of the arch from horizontal at the right-hand side.

If the arch is relatively flat, and θ is relatively small, there is hardly any difference in the cosine of the angle, and the magnitude of compressive forces Fc_1 and Fc_2 is nearly the same. We can simply refer to the compressive force acting on each end of the section as **Fc**.

However, there is a significant difference in the vertical components of the compressive forces. As a result, the *upwards* **vertical component Fc $\sin\theta_1$ is larger than the downwards component Fc $\sin\theta_2$. *These vertical force components do not balance*, since θ_1 is greater than θ_2. The compressive force within the arch generates a net upwards force of **Fc $(\sin\theta_1 - \sin\theta_2)$**.

We have assumed that the arch is relatively flat, so angle θ will be relatively small. Once again, we can use the mathematical simplification that $\sin\theta$ is nearly equal to θ for small angles. Substituting angle θ for $\sin\theta$, we see that the *opposing compressive forces give rise to a net upwards force equal to F $\Delta\theta$*, where $\Delta\theta$ is the change in angle along the section of the arch.

In an arch which is a thin section of a circle, then angle θ varies at a constant rate along the length of the arc. It turns out that the change in angle $\Delta\theta$ along a short segment of length ΔL

is $(8H_{max}/L^2)\Delta L$, so the net upwards force on an arc segment of length ΔL is $Fc(8H_{max}/L^2)\Delta L$. In terms of the compressive force per unit area, **Pc**, the net upwards force on the arc segment is $PcA(8H_{max}/L^2)\Delta L$.

This net upwards force supports the weight of that segment of the arc. The volume of material in the arc segment is its cross-sectional area **A** times the arc length ΔL. If the density of the material in the arch is ρ, then the weight of the material in the segment is $\rho A \Delta L g$.

The net upwards force produced by opposing compressive forces acting on the ends of the arch segment must be equal to the weight of the arch segment. Setting these two terms equal, we get the equation:

$$Pc\cancel{A}\ (8H_{max}/L^2)\ \cancel{\Delta L}\ =\ \rho\cancel{A}\cancel{\Delta L}g$$

Net upwards force produced by opposing compressive forces	Downwards weight of arc segment of length ΔL

Simplifying the equation, and re-arranging the equation, gives us the compressive force **Pc** along each section of the arch.

$$\text{Compressive force Pc} = \frac{1}{8}\,\rho\,g\,\frac{L^2}{H_{max}}$$

This is exactly the same result derived in Equation (3).

Note 1

Using simple triginometry, we can calculate the radius **R** of an arch in terms of its span **L** and its maximum height **H**$_{max}$ mid-way along the arch.

Consider an arch comprising a segment of a circle, as shown in this figure. Let's look at the left side of the arch and, in particular, let's consider the triangle shown shaded in grey (formed by two lines radiating from the centre of the circle to one end and the centre of the arch).

One side of the triangle is half the span of the arch (**L/2**). Another side is distance **X**, equal to radius **R** minus distance **H**$_{max}$. The hypotenuse of the triangle is equal to radius **R**.

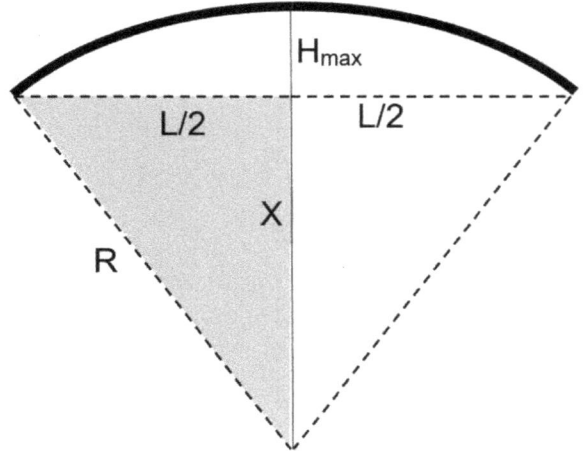

Since this is a right triangle (with a 90 degree angle), we can apply the Pythagorean Theorem. This gives:

$$X^2 + (L/2)^2 = R^2$$

We can re-arrange this equation to get the following expression for distance **X**:

$$X = R \sqrt{\left(1 - \frac{L^2}{4R^2}\right)}$$

Since we are considering arches that are only a small part of a circle, the span **L** is smaller than radius **R**. This means that the term **L²/4R²** is much less than one, and we can apply a useful mathematical simplification to get rid of the square root.

If we consider any number **n** which is much smaller than one, then:

$$\sqrt{1-n} = 1 - \frac{n}{2} \quad \text{When } n \text{ is small compared to 1}$$

You should pick some values of **n** to try this out for yourself. It is 99% accurate even for **n** as large as 0.25, which is really quite astonishing.

We can apply this simplification to the equation for **X** above, and then determine the value of **H**$_{max}$ as follows:

$$H_{max} = R - X$$

$$H_{max} = R - R\left[1 - \tfrac{1}{2}\left(\frac{L^2}{4R^2}\right)\right]$$

$$H_{max} = R - R + \tfrac{1}{2}R\left(\frac{L^2}{4R^2}\right)$$

$$H_{max} = \frac{L^2}{8R}$$

We can re-arrange this result to give Equation (2):

$$\text{Radius of curvature of arch} = \frac{1}{8}\frac{L^2}{H_{max}}$$

116

23. Steel beams: A great innovation of the 19th century

Beams are used to span horizontal distances, and solid wooden beams have been used in roofs, bridges and other structures for hundreds (probably thousands) of years. However, in the 19th century, steel became widely available at relatively low cost, and this was a huge "game changer".

As discussed in the previous chapter, ceramic materials like stone, cement, mortar and concrete were extensively used for monuments and structures in pre-industrial times. These materials provide satisfactory compressive strength to support the weight of large structures, but their tensile strength (ability to resist stretching forces) is much less, making them nearly useless for spanning large horizontal distances.

The mismatch between the compressive and tensile strengths of ceramics contrasts with the properties of iron, steel and other metals. The tensile and compressive strengths of metals are comparable, and are much higher than in most ceramics. Also, the "shear strength" of metals (their ability to resist opposing scissor-like forces) is very similar to their tensile and compressive strengths.

Steel, in particular, provides higher compressive strength than ceramics, and vastly greater tensile strength. Availability of steel in large quantities, consistent quality and low cost greatly extended the ability of engineers to design structures spanning much larger distances and supporting much greater weights (loads). Steel could also be readily fabricated by extrusion and rolling into structural beams of virtually any desired length or shape. Consequently, virtually all large buildings and structures are now made with steel structural beams.

Beams are the simplest means of spanning horizontal distances for floors, roofs and walkways. Wooden beams are usually solid rectangular pieces of timber, although steel beams are generally made of steel sheets that have been folded or extruded into hollow shapes.

As it turns out, the top and bottom of a beam play the most critical role in supporting load forces. The material in the centre of the beam doesn't support much of the load, and mostly acts as a spacer between the material at the top and bottom of a beam. To understand why, imagine that we take a beam with a rectangular cross-section (of width **W** and height **H**), spanning distance **L**. While the beam is initially straight, it will bend if we apply a load force to the beam. The curvature may be imperceptible, but is crucial to how a beam actually supports loads.

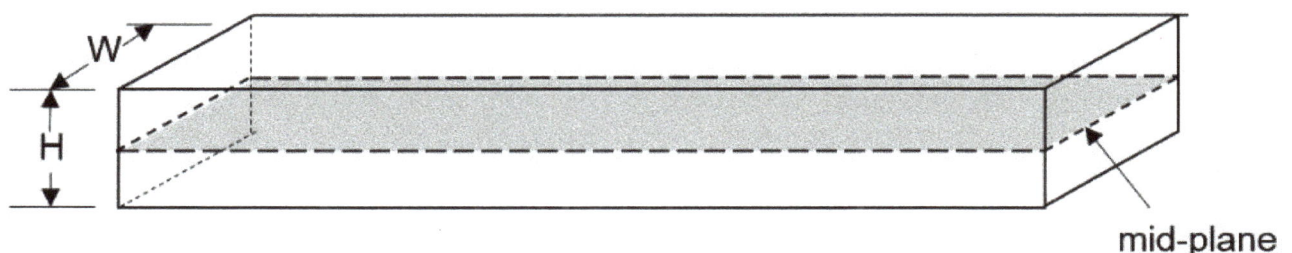

mid-plane

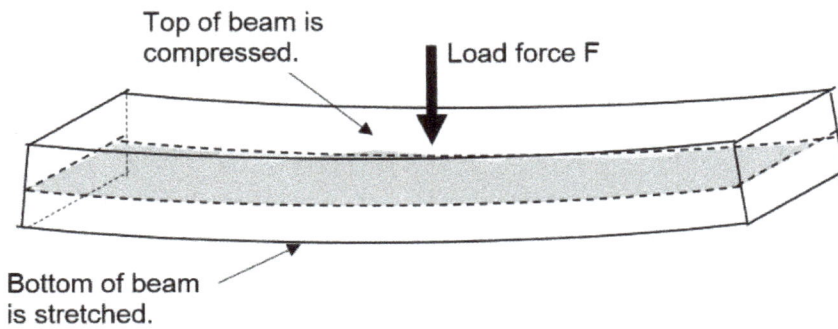

Top of beam is compressed.

Load force F

Bottom of beam is stretched.

Imagine that we apply a downwards load force **F** to the centre of the beam. The beam will curve downwards with a slight circular arc. Since the length along a circular arc increases with the distance from the centre of an arc, the bottom of the beam (at a greater radius) is stretched, while the top of the beam is compressed. There is no change in length along the mid-plane (or "neutral plane" of the beam. Consequently, above the mid-plane of the beam, the material is compressed, and below the mid-plane of the beam, the material is stretched.

The greater the distance above the mid-plane, the more the material is compressed, and the higher the compressive stress. Similarly, the greater the distance below the mid-plane, the greater is the tensile stress within the material. If the load capacity of the beam is exceeded, *the beam will fracture when either the tensile strength of the material is exceeded at the bottom of the beam, or when the compressive strength of the material is exceeded at the top of the beam*. The ability of the beam to withstand loads is mainly determined by what happens at the top and bottom of the beam.

Consequently, it has become common to use "I-beams", which have a cross-section like a capital letter "I". This maximises the cross-sectional area at the top and bottom of the beam, at the greatest distance from the central plane. The material in the central section (which keeps the top and bottom sections apart at distance **H**) can be quite thin. An I-beam can support heavier loads than a solid beam *containing the same amount of material*.

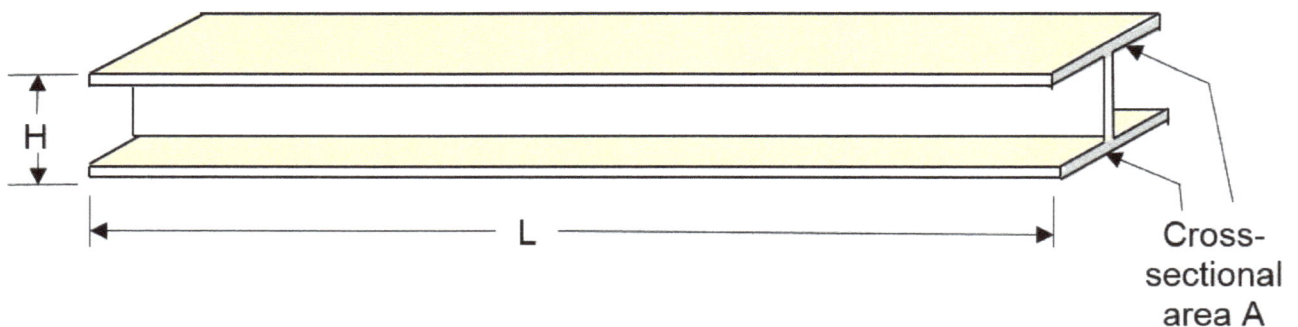

H

L

Cross-sectional area A

An I-beam achieves maximum load-bearing ability with the minimum amount of steel when the top, bottom and central sections of the I-beam are thin (allowing the beam to have maximum height **H** and width **W**). In this case, the load-bearing ability of an I-beam can be as much as three times more than a solid beam with the same amount of material. Load-bearing ability per unit weight is very important in the construction of large structures, since first and foremost, a structure must be able to support its own weight. For very large structures, the weight of the structure itself can be limiting factor determining how large it can be.

We can understand exactly how an I-beam "works", and how much weight it can support, by applying fundamental Laws of Physics that were discovered by Isaac Newton around 1680. Firstly, Newton discovered that when any object (Object A) exerts a force on another (Object

B), Object B exerts the same force on Object A in the opposite direction. So, for example, due to the gravitational attraction of my body to the Earth, my feet exert a downwards force on the floor on which I am standing. The floor exerts the same force, pressing upwards against my feet.

Newton's Second Law states that an object experiencing a **net** force will undergo acceleration (that is, it will increase in speed in the direction that the force is applied). The converse is also true: **If a body is _not_ accelerating, it is _not_ experiencing a net force.** Clearly, we don't want or expect buildings and bridges to move (relative to the Earth), and certainly not to accelerate. So, this means that **there must be no net force acting on such structures, or any part of such structures.** We can even go further. We can infer from Newton's Laws that there must be no net force acting on a structure **in any direction** (horizontally, vertically or diagonally). Of course, there are forces acting on such structures – but these forces must oppose and balance each other.

In the case of a horizontal I-beam with a load force **F** applied at its centre, the vertical downwards force must be balanced by an upwards force exerted by the support at each end. Since this configuration is symmetrical, each support provides an equal upwards force of **F/2**.

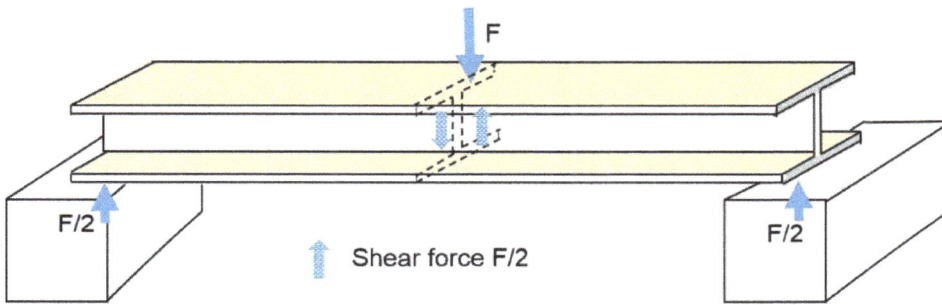

Furthermore, since we know that there is no net force on **any part** of the beam, we can determine the forces acting **within** the beam. Consider the segment extending from its left-hand outer edge to an imaginary surface near the centre of the beam, just to the left of where load force **F** is applied. This section of the beam experiences an upwards force of **F/2** at the left support, and this must be matched by an equal downwards force. However, this downwards force is not provided (not directly, anyway) by load force **F**, which acts on the right-hand section of the beam. Instead, the right-hand section of the beam must exert a downwards shear force **F/2** over the cross-sectional area between the two beam sections. This shear force ensures that all vertical forces on the left-hand section are balanced (Note 1). Furthermore, there are no horizontal forces pushing sideways on the beam.

So far, all the forces are balanced, and everything looks fine. But there is something still missing. Look carefully at the left-hand beam section in the diagram above. There is an upwards force **F/2** acting on its left-hand side, and a downwards force **F/2** acting on its right-hand side. If we cut the beam near the centre, along the imaginary end of the left-hand beam segment, these upwards and downwards forces on opposite ends of the beam section would cause it to rotate like a propeller in the clockwise direction.

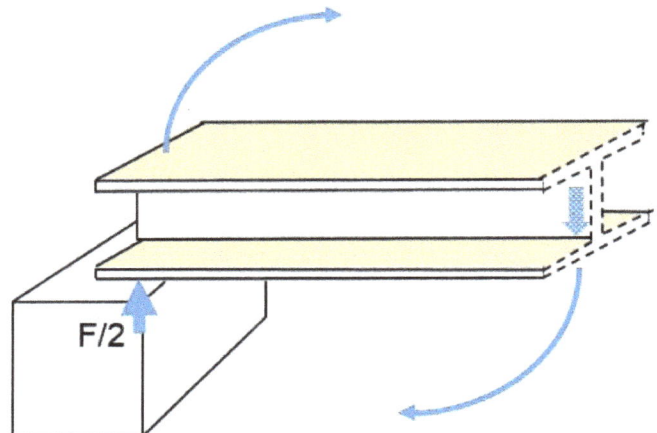

In addition to Newton's Laws, stated above, another condition must apply to every part of the structure. Opposing forces that are displaced by a sideways distance will create a torque, causing the object to rotate. Since we don't want or expect a building or bridge – or any part of it – to rotate, **there must be no net torque operating on any part of the structure**.

Let me explain this with an example of which you are probably very familiar from your childhood. When I was growing up, most playgrounds had a "see-saw", which was basically a wooden plank supported in the middle by a hinged joint, around which the plank could pivot. One child sat on one end and another child sat on the other end. Since the weight of each child acted at the same distance from the hinge, the rotational torque (force times distance from the hinge) exerted by one child was balanced by the opposing rotational torque of the other child. The see-saw was balanced. Of course, as soon as one child pushed against the ground, the plank would rotate and carry him upwards, until the other child pushed against the ground and reversed the rotation.

This works fine if both children have equal weights. However, I was a skinny kid, and usually found myself on a see-saw with an older and heavier kid, who might weigh one-and-a-half times as much as I did. His end of the see-saw would immediately swing to the ground, lifting me up and leaving me suspended in the air with my feet dangling. The remedy was for the heavier kid to sit closer in towards the central hinge.

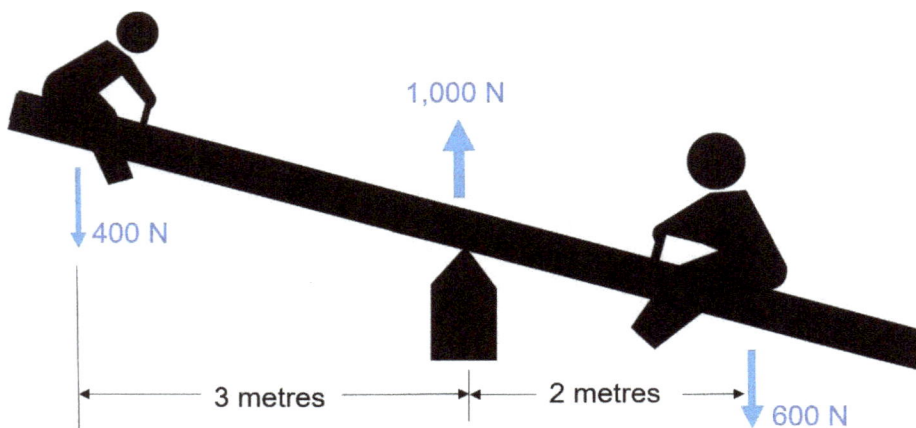

Let's say that my mass was 40 kilograms (a weight of 400 Newtons), and the other kid had a weight of 600 Newtons. If I sat 3 metres from the central pivot point, then my counter-clockwise torque would be (400 Newtons)(3 metres), or 1,200 Newton-metres. If the other kid sat 2 metres from the pivot point, then his clockwise torque would be 1,200 Newton-metres, and there would be no rotation unless one of us pushed against the ground.

In this case, the central hinge of the see-saw is an obvious centre of rotation for calculating the rotational torque, but we could have picked any other point. **If any system is to be stable, there must be no net torque around <u>any</u> centre of rotation that we might choose**.

You might wish to satisfy yourself that there is no net torque around any centre of rotation for the see-saw example given here. You might choose, for example, a centre of rotation at the point where the lighter (or the heavier) child is sitting on the see-saw. Don't forget that the total downwards force of the weight of both children is balanced by an equal upwards force at the pivot.

Coming back to the I-beam, let's consider again the left-hand section of the beam, and imagine a centre of rotation at the left-hand edge of the beam along its central plane. Upwards force **F/2** exerted by the support is directly underneath the pivot point, so it exerts no rotational torque. On the other hand, downwards shear force **F/2** acts at a distance of half the beam length (**L/2**) from the pivot point. This shear force creates a torque of **(F/2)(L/2)** in the clockwise direction.

To balance this clockwise torque, there must be a counter-clockwise rotational torque on the beam section. This is provided by compressive forces in the top of the beam, acting at a distance **H/2** above the mid-plane of the beam, and by tensile forces in the bottom of the beam, acting at the same distance **H/2** below the mid-plane of the beam.

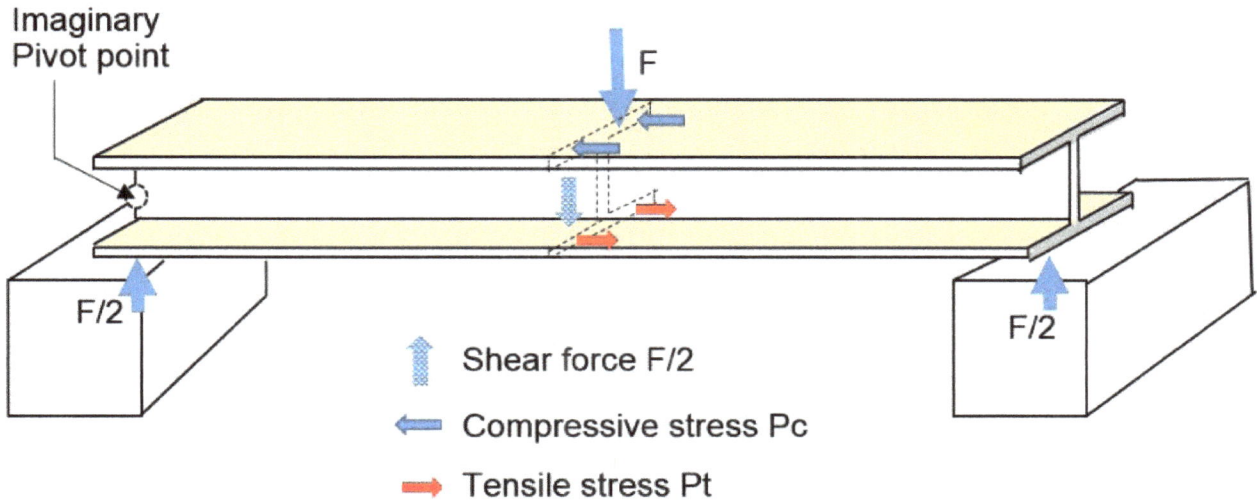

Imaginary Pivot point

F

F/2

F/2

↑ Shear force F/2

← Compressive stress Pc

→ Tensile stress Pt

The compressive stress **Pc** in the top of the beam acts over area **A**, for a compressive force of **PcA**. This compressive force gives rise to a counter-clockwise rotational torque of **(PcA)(H/2)**. The compressive force acting on the top of the beam must be exactly equal to the tensile force acting on the bottom of the beam, so there is no net sideways force on the beam. The tensile force also acts at distance **H/2** from the central plane of the beam, also contributing a counter-clockwise rotation of **(PtA)(H/2)**.

Setting the clockwise torque (due to the shear force acting at the centre of the beam) equal to the counter-clockwise torque (due to the compressive and tensile forces acting on the top and bottom of the beam) gives:

$$(F/2)(L/2) \;=\; (Pc\,A)(H/2) \;+\; (Pt\,A)(H/2) \quad \text{Where } Pc = Pt = P$$

Clockwise torque of shear force

Counter-clockwise torque of tensile & compressive forces

So, $\tfrac{1}{4}\,F\,L = P\,A\,H$

> Where **F** is the load force applied at the centre of the beam
> **L** is the length of the beam
> **P** is the tensile stress in the bottom of the beam and the compressive stress at the top of the beam.
> **H** is the height of the beam

Usually, a beam will fail when the tensile stress at the bottom of the beam exceeds its tensile strength Pt_o. Simplifying and re-arranging the equation allows us to calculate the maximum load force **F** that can be supported by the beam:

$$\text{Maximum load force } F_{max} = \frac{4\,Pt_o\,A\,H}{L}$$

This equation tells us, as we might expect, that the load that can be carried by a beam increases with the tensile strength of the material, the height and areas of the top and bottom sections of the beam, and reduces as the beam gets longer.

Alternatively, if we need to support a given load force **F** in the middle of a span of length **L**, we need to know the minimum vertical height of the beam. Re-arranging the equation above gives:

$$\text{Minimum beam height, } H_{min} = \frac{F\,L}{4\,Pt_o\,A}$$

As we might expect, the minimum height of the beam increases with the applied load force and the length of the beam, but varies inversely with the tensile strength of the material.

These equations relate the key parameters for an I-beam with a load force acting at a single point. However, in most "real world" situations, we are dealing with "distributed loads" in which load forces are spread along the length of the beam. If we are dealing with a long beam, the weight of the material in the beam is a significant load, and is spread evenly along its length. By applying the same reasoning, we will be able to derive a similar equation relating the key parameters and performance of an I-beam with a distributed load. Furthermore, we will be able to calculate the maximum possible length at which a concrete or steel beam will break under its own weight.

Notes
(1) Simultaneously, of course, the left-hand section exerts an upward shear force **F/2** on the right-hand section, so shear forces exert no net force on the entire beam.

24. Distributed loads on beams and trusses

Previously, we had considered the effect of a load force applied at a single point to a beam. In most cases, however, the load force on a beam or truss is applied across its entire length.

Let's consider the case of a load force that is applied uniformly along a beam. This would include, of course, the weight of the beam itself. In the case of a bridge, the structure must support its own weight plus the weight of vehicles or pedestrians travelling along its length. A roof truss supports the weight of ceiling tiles along the entire roof area.

Once again, we'll consider an ideal I-Beam, in which nearly all of the material in the beam is located along its top and bottom surface (where maximum compression and stretching of the beam occurs, and the resulting compression and tensile forces produce the greatest torque to counter-act the load forces). The beam is of length **L** and height **H**, and has cross-sectional area **A** at the top and bottom of the beam. Let's now consider a downwards load force acting over the entire length of the beam, with a constant force **f** per unit length.

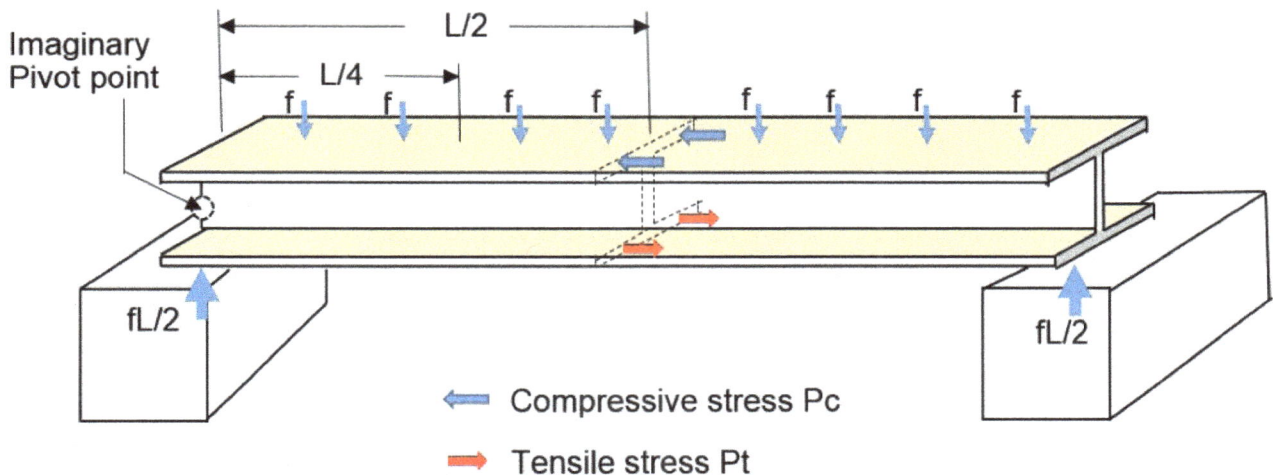

Since the beam is not moving or accelerating, there must be no net force acting on the beam in the vertical direction (or in any other direction). Since force **f** acts on each metre along the length of the beam, and the beam is **L** metres long, the total load acting on the beam is **fL**. This must be balanced by upward forces exerted by the two supports at the ends of the beam. Hence, the upwards force exerted by the support at each end of the beam is **½fL**.

Once again, let's focus our attention on the left-hand side of the beam, starting at the left-hand end and going to the mid-point of the beam. Since the length of this section is **L/2**, the downwards load force acting on this section of the beam is **fL/2**, which is exactly equal to the upwards force **fL/2** exerted by the support on the left-hand side. Consequently, in this case, there is no shear force acting at the centre of the beam (although we can easily show that there is a shear force that increases towards the ends of the beam).

123

Although the downwards and upwards forces acting on the beam are balanced, they still generate a torque that must be opposed if the beam section is not to spin. Once again, it is convenient to consider an imaginary pivot point at the mid-plane on the left-hand edge of the beam.

As before, upwards force **fL/2** exerted by the support is directly underneath the pivot point, so it exerts no rotational torque. On the other hand, the uniformly-distributed load force **f** acts along the entire section of the beam, from directly above the pivot point up to distance **L/2** from the pivot point. If we add up the torque contribution by the load forces acting along the beam section, the load force **fL/2** acts at ***an average distance of L/4*** (halfway along the left-hand section of the beam). Consequently, the uniformly-distributed load force produces a torque of **(fL/2)(L/4)**, or **fL²/8,** in the clockwise direction. This is half as much torque as would be produced if the total load force acted at the centre of the beam.

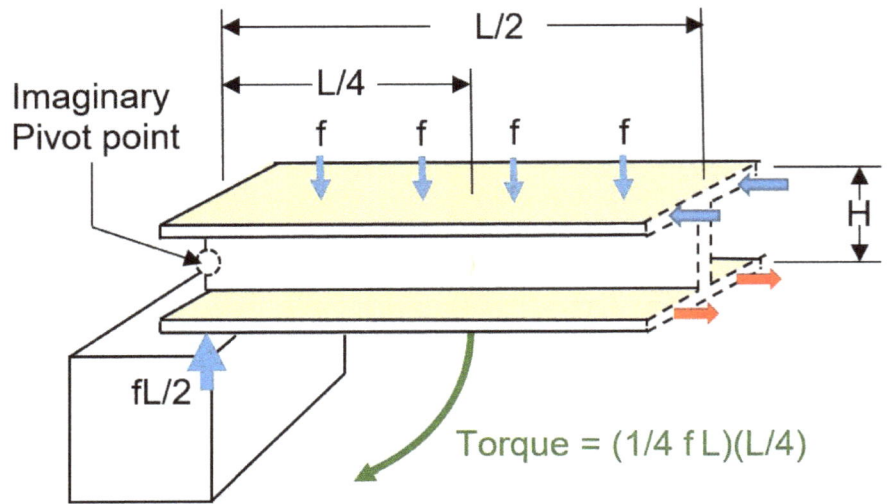

Once again, the clockwise torque produced by the load force is opposed by a counter-clockwise torque produced by the compressive force acting on the top of the beam segment (at height **H/2** above the pivot point) and the tensile force acting on the bottom of the beam segment (at height **H/2** below the pivot point).

At the top of the beam, compressive stress **Pc** acts over area **A**, for a compressive force of **PcA**. This compressive force acts at height **H/2** above the mid-plane of the beam (and above the pivot point), giving rise to a counter-clockwise rotational torque of **(PcA)(H/2)**.

The compressive force acting on the top of the beam must be exactly equal to the tensile force acting on the bottom of the beam (otherwise, there would be a net force pushing the beam sideways). The tensile force also acts at distance **H/2** from the central plane of the beam, and also contributes a counter-clockwise torque of **(PtA)(H/2)**.

Setting the clockwise torque exerted by the load force equal to the counter-clockwise torque exerted by the compressive and tensile forces acting on the top and bottom of the beam) gives:

$$fL^2/8 = (PcA)(H/2) + (PtA)(H/2) \qquad \text{Where Pc = Pt}$$

When the maximum load force is applied, the beam will break and collapse, either because the compressive or tensile strength of the material is exceeded. Let's say that the maximum load that a beam can support is limited by the tensile strength of the material Pt_o,

By simplifying and re-arranging the equation above, we can derive the maximum distributed load force that a beam can support before it fails.

$$\text{Equation (1)} \qquad f_{maximum} = \frac{8\, Pt_o\, A\, H}{L^2}$$

Where $f_{maximum}$ is the maximum load force per metre length of the beam
Pt_o is the tensile strength of the material in the beam.
A is the cross-sectional area at the top of the beam, and also at the bottom of the beam.
H is the height of the beam
L is the length of the beam

Perhaps a more interesting result is to use this equation to derive the maximum possible length of an I-Beam that can support its own weight. In this case, the uniformly-distributed load is simply the weight per unit length of the beam.

The mass of the beam is simply the volume of material in the beam times its density **ρ**. The volume of material in the beam is the cross-sectional area **2A** (taking into account material at both the top and bottom of the beam) multiplied by length **L**. We multiply the mass of the beam by the acceleration of gravity **g** to get its total weight, **ρ (2AL) g**. This is a simplification, as we are neglecting the vertical sheet of material separating the top and bottom of the beam, so the result will understate the actual weight. To get the weight per unit length, we divide the total weight **ρ (2AL) g** by length **L**.

$$\text{Weight per unit length, } f = \frac{\rho[(2A)\cancel{L}]g}{\cancel{L}}$$

$$\text{Equation (2)} \qquad \text{Weight per unit length } f = 2\,\rho\,A\,g$$

We can insert equation (2) into Equation (1) to derive the maximum length of a beam that can just support its own weight. We get:

$$2\rho\cancel{A}g = \frac{8\, Pt_o\, \cancel{A}\, H}{L^2}$$

Which simplifies and re-arranges to:

$$\text{Equation (3)} \qquad \text{Maximum possible length } L = 2\sqrt{\frac{Pt_o\, H}{\rho\, g}}$$

Consider the maximum possible length of a beam of unreinforced concrete. The limiting factor is the tensile strength of the concrete Pt_o, which is about 3 megapascals (3 X 10^6 Newtons per square metre). Concrete has a density ρ of about 2,300 kilograms per cubic metre. The acceleration of gravity on Earth, **g**, is about 10 metres/second^2. Inserting these figures, gives:

$$\text{Maximum possible length of concrete beam} = 23\,\sqrt{H}$$

Thus, an unreinforced concrete beam with a height of one metre, which is a very substantial beam, would have a maximum possible length of about 23 metres. This is an absolute maximum possible length. Such a beam would just barely be able to support its own weight, and would collapse if any other load were applied (say if a bird landed on it). Furthermore, we have neglected the weight of the central section of the beam, and assumed that the beam has no cracks or flaws. We have ignored potential additional forces arising from high winds during storms, impacts, earth tremors, etc. (called "dynamic loads").

Of course, we could increase the maximum possible length of the beam by simply increasing its height, and indeed, this is the main strategy used to achieve long spans. This is the idea behind a truss, in which the top and bottom of the beam are replaced by one (or more) girders extending the full length of the truss. The top and bottom girders are maintained at separation **H** by a series of shorter girders linking the top and bottom, often in a triangular pattern. These oppose the vertical and sideways shear forces exerted by the top and bottom girders (equivalent to the role played by the central section of an I-Beam).

Nowadays, probably no engineer in his (or her) right mind would use unreinforced concrete beams to span significant distances. So, let's consider the question "What minimum height is needed for a steel beam or truss to span a given distance?"

The answer is given by re-arranging Equation (3), to give:

Equation (4)   Minimum height required, $H_{minimum} = \dfrac{\rho \, g \, L^2}{4 \, Pt_o}$

Bear in mind that this is an absolute minimum for a beam or truss that is just about to collapse. To achieve a satisfactory margin of safety, we would make the beam or truss several times as high.

A typical steel would have tensile and compressive strengths of about 1,000 megapascals, and a density of about 7,800 kilograms per cubic metre.

Let's consider the design of the Oshima Bridge in Yamaguchi Prefecture, Japan. This bridge is basically a simple steel truss, with a central section spanning 325 metres. This the 5th longest truss span in the world.

From Equation (4), we derive that a steel truss spanning 325 metres must have a minimum height of two metres to just support its own weight. The actual height of the central span of the Oshima Bridge is about ten times this minimum figure.

Author/Source: Kiensvav & Glabb/Wikimedia Commons
http://structurae.net.photos/251

Once you are aware of beams and trusses, and the vital role they play, you notice them everywhere (the same applies for many things).

It might seem that we can extend a beam or truss to any length we would like, simply by increasing its height. Note however, that increasing the height of a beam or truss follows a "diminishing law of returns". To make the beam twice as long, we would have to increase its height by four times. And, for very long spans, Equation (4) no longer works. We have assumed that the central section of an I-Beam or truss (holding the top and bottom sections

A truss forms the central section of the Story Bridge crossing the Brisbane River.

at the correct spacing) can be very thin and light. But, as the length of a beam or truss increases, the central section is subjected to increasing shear, tensile and compressive forces.

To span very large distances, different design strategies are used, although the same principles apply. We'll consider these types of designs in the next chapters, and finish up by determining the maximum possible distance that can be spanned by *any* type of structure.

25. Cable-stayed bridges

One of the major types of bridge construction in recent times has been the cable-stayed bridge, which has superseded the suspension bridge in many new constructions.

As a general rule for bridge construction [Reference 1]:

- For short-span bridges, the most practical and cost-effective design strategy is to use a truss.

- For medium-span bridges, cable-stayed designs are often the preferred option.

- For bridges of the longest spans, suspension bridges are (still) the preferred option.

There are many exceptions to this general rule, and cable-stayed bridges seem to have broadened their range into applications that traditionally would have been served by truss or suspension bridges.

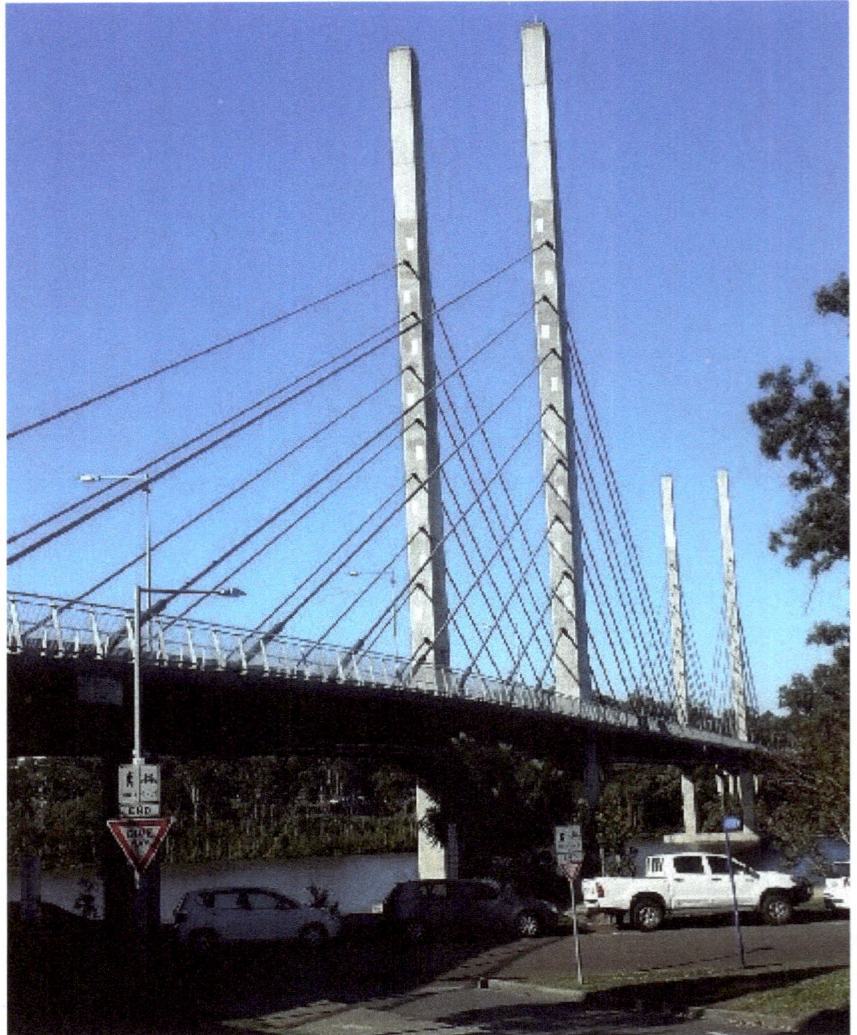

Some of the longest bridge spans in the world are cable-stayed bridges. The Russky Bridge (Russia), Sutong Bridge (China), and Stonecutters Bridge (Hong Kong) have spans exceeding one kilometre. Some smaller bridges (such as the Eleanor Schonell bus/pedestrian bridge at the University of Queensland, pictured here) also use cable-stayed design.

The cable-stayed bridge uses tall towers to support cables, which fan out to the roadway. Tension forces in these cables support the weight of the bridge roadway. Understanding how cable-stayed bridges are able to span long distances is an interesting physics problem, and can readily be done using the same simple laws that we applied previously to beams and trusses.

A cable-stayed bridge basically consists of a horizontal roadway, with towers spaced periodically along its length. Let's say that each tower rises distance **H** above the roadway. A series of cables (normally made of steel) are anchored in the roadway, rise diagonally to a point on the tower, and then descend to the roadway on the opposite side of the tower. In a nutshell, a large span is achieved with cable-stayed bridges by constructing high towers (while, a large span can only be achieved by a truss which is very high along its entire length).

The Eleanor Schonell Bridge at the University of Queensland is a "harp-style" cable stayed bridge, in which the cables run essentially parallel. Cables from the centre of the roadway intersect the tower at the highest point, while cables from near the base of the tower intersect lower down. Consequently, all the cables intersect the roadway at the same angle. For each cable, the ratio of the distance from the base of the tower to the height reached on the tower is the same. This "harp pattern" is commonly used, even though it leads to higher compressive stress in the bridge roadway than for cable stayed bridges in which the cables go to the top of the tower ("radial pattern").

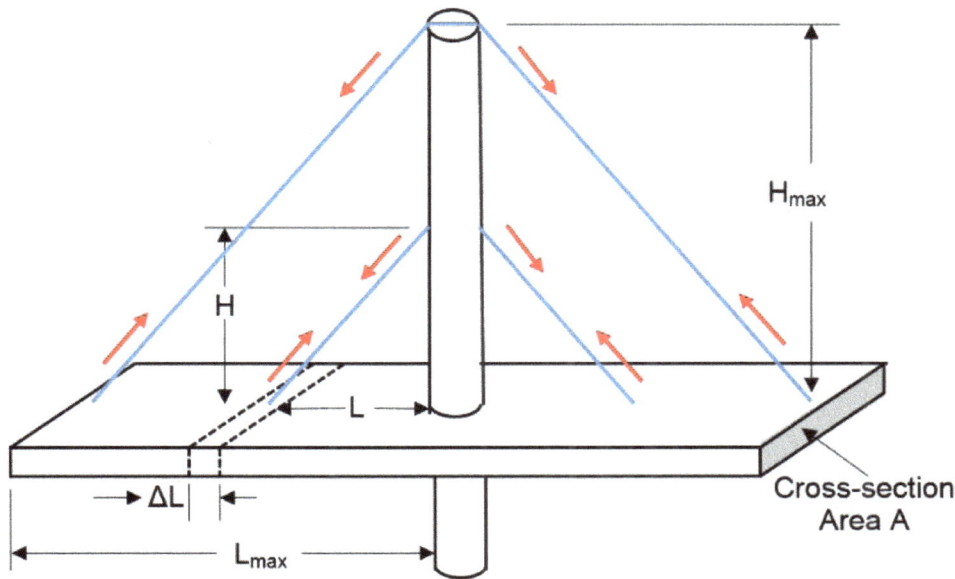

In all cable stayed bridges, the cables support the weight of the roadway, which exerts a tension force on the cables. Normally, cables splay out symmetrically from the tower, so the sideways (horizontal) pull exerted by each cable on the tower is balanced by an equal and opposite sideways force exerted by a cable on the opposite side of the tower. The tensile force in the cables also exerts a downwards (vertical) force on the tower, and downwards forces from all the cables leading to the tower are additive, placing the tower in compression.

Consider a section of roadway (at distance L from the tower), from which one cable extends to the top of the tower. The tensile force along the cable is directed diagonally (at angle θ from vertical). This tension force exerts a sideways (horizontal) force component towards the tower, and an upwards (vertical) force component on the roadway. Each cable supports a short section of roadway of length ΔL. Let's assume for now that the roadway has a constant cross-sectional area A along the length of the bridge span.

Let's take a close look at the tension force along the cable where it intersects the bridge deck at an angle θ. This angle is determined by the height H of the tower and the distance L that the cable intersects the bridge deck from the tower.

The tension force F has two components:

- A vertical, upwards force F_v, lifting the bridge section. Of course, for the bridge to be stable, there must be no net vertical force on the bridge section, so the upwards force F_v is exactly equal to the downwards weight of the bridge section.

- A horizontal force F_h, pulling the bridge section towards the tower.

Forces on segment of roadway

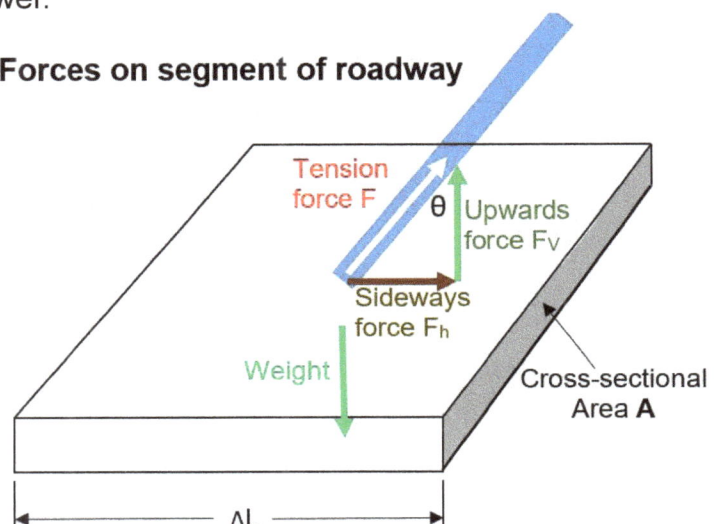

The role of the cable is to provide the vertical force component to hold up the section of the bridge roadway. We don't need or want the horizontal force component, but it is unavoidable due to the angle of the cable.

The vertical and horizontal force components F_v and F_h are in the same proportion as the vertical and horizontal sides of the triangle formed by the cable. With the "harp pattern", the ratio of the vertical and horizontal sides of the triangle is the same, given by height H_{max} and length L_{max}.

Equation (1) $$\frac{F_h}{F_v} = \frac{L_{max}}{H_{max}}$$

To provide the upwards component F_v needed to hold up the bridge section, while minimising the sideways force component F_h, we should make the towers as tall as possible (relative to the distance spanned **L**).

Let's consider how tall we need to make the towers for the bridge to span a given distance.

Let's apply Equation (1) to our bridge section of length **ΔL** and cross-sectional area **A**. We know that the vertical force component F_v must be equal to the weight of the bridge section. The mass of material in this section is its volume **AΔL** multiplied by the density of the material **ρ**. The weight of the bridge section is **ρ(AΔL)g**. Consequently:

$$F_v = \rho \, (A \, \Delta L) \, g$$

Inserting the value of F_v into Equation (1) tells us the horizontal force component exerted on the bridge section by each section of length ΔL:

Equation (2) $$F_h = \frac{L_{max}}{H_{max}} \, [\rho \, (A \, \Delta L) \, g]$$

Each section of length **ΔL** contributes the same sideways force component F_h, pushing the entire roadway towards the tower. The cables are equally spaced so each section pushes on the next section, which pushes on the next. The sideways force F_h produced by **each** section adds to the forces produced by roadway sections that are further from the tower. The effect is similar to a row of people, each one pushing the next and contributing to the total force applied at the end of the line.

The compressive force in the roadway reaches a maximum at the bridge pylon.

$$F_{max} = F_h = \frac{L_{max}}{H_{max}} \, [\rho \, A \, g] \, [\Delta L_1 + \Delta L_2 + \Delta L_3 + \Delta L_4 + \ldots .]$$

Where ΔL_1, ΔL_2, ΔL_3… are the lengths of bridge sections 1, 2, 3, and so forth.

Note that the lengths of all the bridge sections simply add up to L_{max}, so the compressive force reaches its maximum value F_{max} at the tower:

Equation (3) $$F_{max} = F_h = \frac{L_{max}}{H_{max}} \, [\rho \, A \, g] \, L_{max}$$

The maximum compressive stress in the roadway P_{max} is the compressive force at the pylon F_{max} divided by the area A of the roadway.

$$P_{max} = \frac{F_{max}}{A} = \frac{\rho \cancel{A} g L_{max}^2}{H_{max} \cancel{A}}$$

The result gives the maximum compressive stress in the roadway (where it contacts the tower):

Equation (4) $$Pc = \frac{\rho g L_{max}^2}{H_{max}}$$

Where ρ is the density of material in the roadway.
g is the acceleration of gravity (9.8 metres/second²)
L_{max} is the length of the roadway from the tower to the centre of the span.
H_{max} is the height at the tower reached by the highest cable.

Let's apply this to the Eleanor Schonell Bridge, which has a main span of 185 metres (L_{max} has half this value), and where the cables intersect the pylon at about 20 metres above the roadway. Substituting the values of L_{max}= 93 metres, H_{max} = 20 metres, and assuming that the roadway is made primarily of steel with a density of 7,800 kilograms per cubic metre, gives a compressive stress of 33 megapascals. This should give a good margin of safety, assuming that the roadway structure is made largely of steel (with compressive strengths of about 1,000 Mpa) and concrete (with compressive strength of 30 Mpa).

If the span of a cable stayed bridge is too long (relative to the tower height), the compressive stress in the roadway at the tower exceeds the compressive strength of the material Pc_o, and the roadway will crush. Equation (4) allows us to determine the maximum span at which the roadway will fail. The maximum total span between towers is $2L_{max}$. By substituting Pc_o for Pc and re-arranging Equation (4), we can find the maximum span between the towers for a "harp-style" cable-stayed bridge.:

Equation (5)   Maximum span for "harp-style" cable stayed bridge $= 2\sqrt{\dfrac{Pc_oH}{\rho g}}$

Where Pc_o is the compressive strength of the material in the roadway

Equation (5) gives exactly the same result (derived in the previous chapter) for the maximum span for a truss, except that:

- The span is limited by the height of the tower, rather than by the height of a truss along its entire length.

- The span is limited by the compressive strength of the roadway, rather than by tensile strength of girders in a truss.

We could reduce the compressive stress in the roadway, and increase the span, by constructing a cable stayed bridge with the "radial pattern" - that is, with all the cables going to the highest possible point on the pylon tower. This allows cables that are closer to the tower to be inclined at a steeper angle, reducing the sideways force component relative to the vertical force holding up each bridge section. So, for example, a roadway segment that is half the

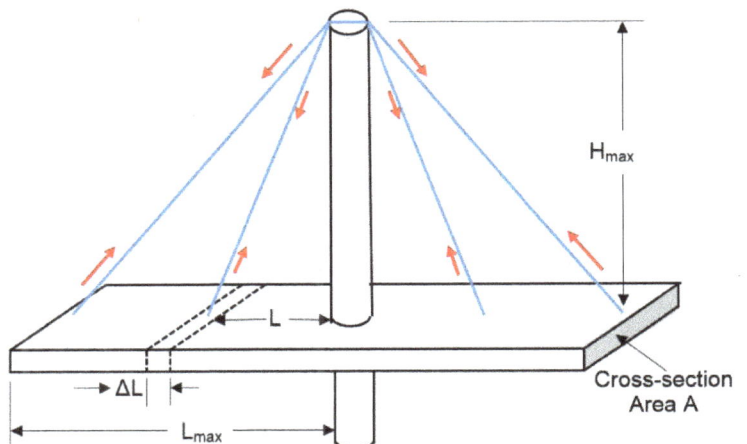

distance from the pylon will produce half the sideways force component. However, although a "radial pattern" is advantageous in reducing compressive stress in the roadway, the pylons may need to be wider (since cables do not provide bracing for the centre of the tower) and some designers or clients prefer the aesthetics of "harp-style" construction.

By using "radial" rather than "harp" construction, the net effect is to reduce the maximum compressive stress by half, and to increase the maximum bridge span by the square root of 2 (or 1.41 times).[Note 1]

Of course, the actual span of the bridge is reduced by the "live loads" it must support (the weight of vehicles and people travelling over the bridge). Usually, the "live load" of a long-span bridge is relatively small, less than 10% of the weight of the bridge structure itself, and so, does not have a large effect on the maximum bridge span.

I have plotted a graph of Equation (4), showing how the compressive stress increases along the roadway, starting at the centre of the span and extending distance **L** to the tower. I have assumed that the roadway is made of steel (with compressive strength **Pc** of 1,000 MPa and density ρ of 7,800 kilograms per cubic metre). I have also assumed a tower height of 10 metres, a cross-sectional area of the roadway of 1.0 square metre, and a "live load" of 10,000 Newtons/ metre (equivalent to two lanes of cars parked bumper-to-bumper). With these assumed values, the roadway will undergo catastrophic compressive failure when the span from tower to tower (distance **2L**) reaches about one kilometre.

Compressive stress in bridge roadway versus span length L

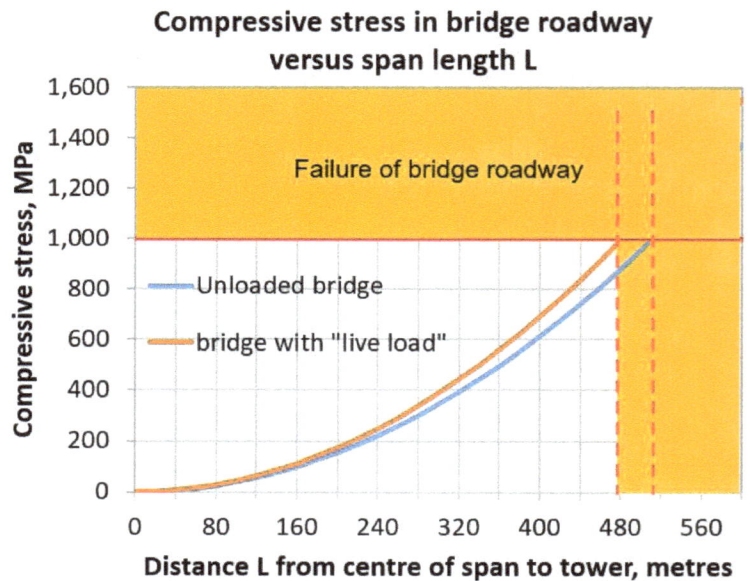

The maximum span can be increased – as we might expect – if we use a material with higher compressive strength and lower density, or if we build higher towers. But these measures reach a point of diminishing returns. For example, to double the span of the bridge, we must increase the height of the towers by four times.

In our calculation of the maximum bridge span given by Equation (5), we have assumed a constant cross-sectional area for the roadway. This provides for a simple design, but does not achieve the maximum bridge span. To understand why, recall that compressive stress in the roadway is relatively small near the centre of the bridge span, with the compressive stress increasing closer to the tower. The roadway near the centre of the span, which is subject to very little compression stress, could have a much smaller area (and less weight) than the roadway near the tower. We could – at least in theory – design the bridge so that the cross-sectional area of the roadway increases from the centre of the span to the tower. The idea is that the cross-sectional area at any distance along the span would be just enough to withstand the compressive force at that point. This would reduce the weight of the roadway, reduce the tensile stress in the cables, and minimise the compressive stress on the roadway.

It is not difficult to calculate the minimum cross-sectional area at each point along the roadway that would just withstand the compressive stress (due to the weight of the roadway itself and the "live load"). It turns out that this cross-sectional area varies exponentially with distance **L** from the centre of the span.

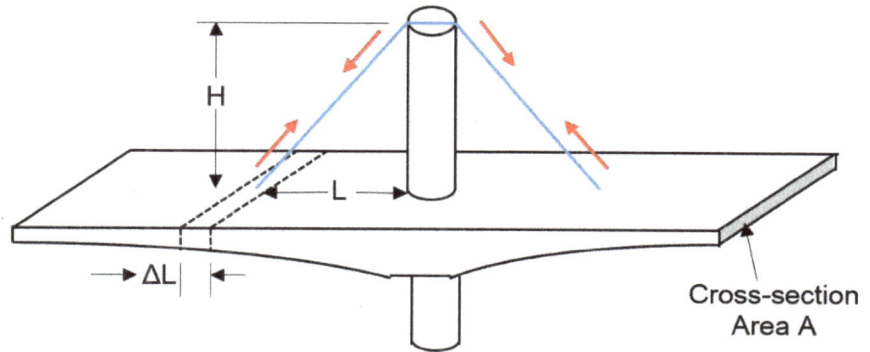

Cross-section Area A

The result that I calculate is as follows [Note 2]:

Equation (6) Minimum cross-sectional area of roadway $= \dfrac{f}{\rho g}\left[e^{\rho g L^2 / 2 P_{co} H} - 1\right]$

Where **f** is the "live load" supported by the bridge (Newtons per metre along the roadway)

ρ is the density of the material in the roadway

Pco is the compressive strength of material in the roadway

H is the height of the towers

L is the length of the roadway from the centre of the span

I have plotted a graph of Equation (6), showing how the cross-sectional area of the roadway would vary with distance from the centre of the span (I have used the same values for density and compressive strength of steel, a "live load" of 10,000 Newtons/ metre, and a tower height of 10 metres). By varying the area of the roadway in this way, we could theoretically increase the bridge span as long as we like. However, while we could perhaps double the span of the

Minimum area of roadway to support bridge

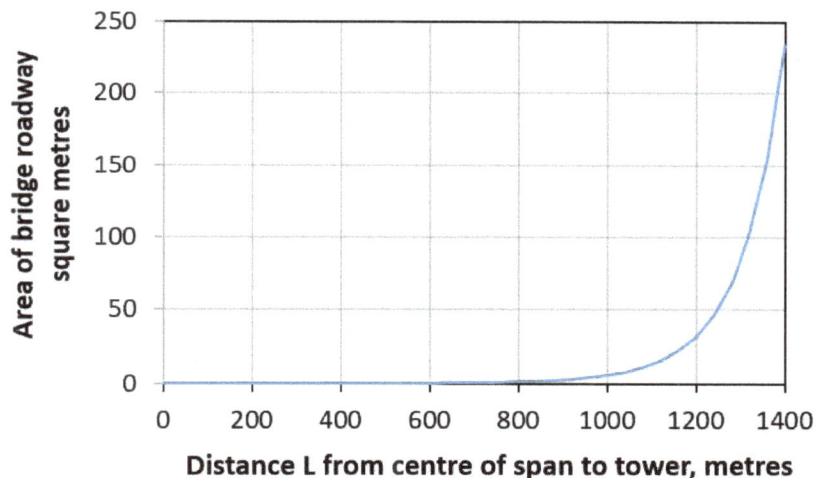

bridge in this way, increasing the span beyond that becomes completely impractical. Once the span of the bridge exceeds the maximum value calculated previously, the required roadway area simply explodes exponentially.

Even by varying the area of the roadway or increasing the height of the towers, we reach a limit at which further increases in span become unachievable. In the preceding analysis of cable-stayed bridges, I have only considered the weight of the roadway, and have completely ignored the weight of the steel cables holding up the roadway. I have also ignored the effect of gravity on the cables, which causes them to "sag" (curving downwards in a parabolic curve, rather then being straight). Both of these assumptions are perfectly valid – providing that the span of the bridge is not too great. However, as the span increases, cables become longer (to reach across the wide span) and must be thicker (to provide increased tension forces to compensate for the shallower angle). Consequently, as the span increases, the weight of the cables increases in comparison with the weight of the roadway. At very long spans, the cables account for most of the weight of the bridge! The weight of the cables causes them to curve (rather than be straight). In effect, as we increase the span of a cable-stayed bridge, it begins to morph into a suspension bridge.

References

(1) Bridge Engineering Handbook, 2nd Edition: Fundamentals, Wai Fah Chen and Lian Duan, 2014, See Google Books

Notes

(1) In a cable stayed bridge with cables in a "radial pattern", all the cables originate from the top of the pylon tower at height H_{max}, and intersect the roadway at distance L from the tower. Equation (2) needs to be replaced by:

$$F_h = \frac{L}{H_{max}} [\rho (A \Delta L)g]$$

The sideways force components F_h produced by each segment of length ΔL are additive, so that the maximum compressive force F_{max} at the pylon is the sum of all the sideways force components.

$$Pc = \sum \frac{F_h}{A}$$

If ΔL is small compared to the total length L, then the sum becomes an integral, and the compressive force at the tower becomes:

$$Pc = \sum \frac{L}{H} [\rho (\Delta L)g] = [\frac{\rho g}{H}] \int_{L=Lmax}^{L=0} L \, dL$$

$$Pc = 1/2 \frac{\rho g L_{max}^2}{H}$$

Thus, the compressive force in the roadway for a radial cable pattern is half that for a "harp style" cable pattern.

(2) The starting point to derive Equation (6) is with Equation (2). Taking into account the "live load" f, the horizontal force component contributed by each roadway section of area A and length ΔL is:

$$\Delta F_h = \frac{L}{H} [\rho (A \Delta L) g + f \Delta L]$$

We assume that area A is just enough to resist the compressive force. The force per unit area on the roadway is the compressive strength of the material Pco.
    We substitute $A = F_h/Pco$ into the equation above, to get:

$$\Delta F_h = \frac{L}{H} [\frac{\rho g}{Pco} F_h + f] \Delta L$$

This equation can be solved by moving all terms with ΔF_h and F_h on one side, and all terms with L and ΔL to the other side. We assume that ΔL is much smaller than L, so the equation can be integrated. The compressive force F_h varies from zero (at the centre of the span) to its maximum value at the tower. With some re-arrangement, this leads to Equation (6).

26. Suspension Bridges

Like many cities of the world, New York is bound into a cohesive unit through its bridges. All of its iconic bridges that are well-known around the world are suspension bridges. Most were built during a period of rapid population expansion and technological advancement from the 1870s to the 1920s. This bridge-building boom was enabled by development of technology to extrude steel wires and wind them into steel cables that were long, very strong (tensile strength) and flexible.

This started with construction of the Brooklyn Bridge, spanning the East River from Lower Manhattan to Brooklyn. Within the next few decades, the Williamsburg and Manhattan Bridges were constructed within a few kilometres of each other (and near where I lived when I was growing up).

Then, the George Washington Bridge, Bronx-Whitestone Bridge and Throgs Neck Bridge were completed. These and other suspension bridges have become iconic structures defining the skyline of New York. While I lived in New York, these bridges were an integral part of my life and the lives of most other New Yorkers. I have walked, drove or taken the subway over each of these bridges – many times.

In 1962, on my 14th birthday, My family moved to Brooklyn. From out apartment on the 13th floor, we had a commanding view over Lower New York Bay. From our living room, we watched as the Verrazanno Narrows Bridge was built across the bay between Brooklyn and Staten Island. Its main span is 1,298 metres, making it the longest suspension bridge in the United States. Years later, I would drive over the Verrazano Bridge to my first full-time job in New Jersey.

So, perhaps I have a biased view that suspension bridges are the ultimate bridge technology. While other bridge designs (such as cable-stayed bridges) are often employed in modern bridges with short-to-moderate spans, suspension bridges are still the leading technology for the longest bridge spans. I expect that this will remain the case. At the end of the day, if a span is too long to stretch a cable from one side to the other, then it is too long to build a bridge.

Akashi Kaikyo Bridge
Self-published work by Tysto, CC BY-SA 3.0
https://commons.wikimedia.org/w/index.php?curid=477955

The bridge with the longest span is the Akashi Kaikyō suspension Bridge, which crosses the Akashi Strait from Kobe to Awaji Island in Japan. Its main span is 1,991 metres. The two towers were originally 1,990 metres apart when the bridge was first constructed, but the towers moved apart one metre during an earthquake on January 17, 1995.

Each of the main suspension cables is more than a metre in diameter, and has a mass of more than 15,000 tonnes.

In a suspension bridge, flexible steel cables are slung between towers (or "pylons"). Smaller steel cables are suspended from the main cables, holding up the weight of the roadway. The weight of the roadway and the cables themselves cause the main cables to curve into a circular arc. The curvature of the cables, combined with tensile forces within the cable, give rise the upwards force supporting the weight of the bridge.

The main cables take on the shape of the bottom segment of a circle (which is essentially a parabola).

We had earlier considered the construction of arch bridges, which are essentially a mirror image of a suspension bridge:

- Arches form the top segment of a circle, and *its weight is supported by compressive force within the downwards-curving arch*.

- Suspension cables form the bottom segment of a circle, and *its weight is supported by tension forces within the upwards-curving cable*.

In an arch bridge, the arch must have sufficient compressive strength so that it can withstand the compressive stress to which it is subjected. Arches, and all other structures subject to compressive forces, must also be sufficiently stiff, wide and thick to avoid buckling. But buckling is not a consideration for structures subject to tensile forces. The cable of a suspension bridge must simply have sufficient tensile strength to withstand the tension force to which it is subjected. Cables do not need to be stiff. In fact, the opposite applies. Cables must be flexible, so they can bend to the shape necessary to support its own weight, as well as the weight of the roadway and other "loads".

Verrazano Bridge, Photo by Mike LaMonaca
httpscommons.wikimedia.orgwikiFileNew_York_City_Verrazano-Narrows_Bridge.jpg

Let's consider the forces within a section of cable in a suspension bridge. Let's say that the section of cable has a short length ΔL. This section of cable might support a section of roadway of the same length through a vertical steel cable (called a "suspender" running between the cable and roadway). Since the cable (and this particular section of cable) is not moving and not accelerating, the net force on this section of cable must be zero. Furthermore, the net force *in any direction* acting on this section of cable must also be zero.

136

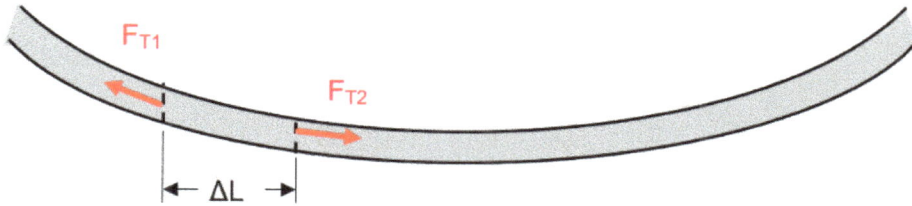

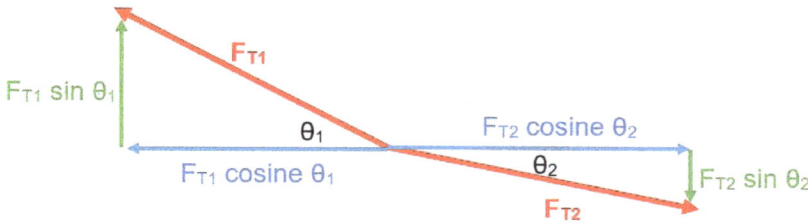

On one side, the cable is pulled to the left and upwards by the tension force in the cable F_{T1}. The other side of the cable is pulled downwards and to the right by tension force F_{T2}. However, because of the curvature of the cable, these forces do not fully cancel each other. The upwards-sloping angle θ_1 of tension force F_{T1} (relative to horizontal) is greater than the downwards-sloping angle θ_2 of tension force F_{T2}.

We can resolve each of the tension forces, F_{T1} and F_{T2}, into horizontal and vertical components. The horizontal component of tension force F_{T1} is this force multiplied by the cosine of angle θ_1 (the ratio of the adjacent side of the triangle to its hypotenuse). Similarly, the horizontal component of tension force F_{T2} is this force multiplied by the cosine of angle θ_2. As it turns out, the cosine of an angle hardly changes until the angle begins to exceed 20-30 degrees.

The two opposing horizontal forces, $F_{T1} \cos \theta_1$ and $F_{T2} \cos \theta_2$ are the only forces acting in the horizontal direction, so they must add to zero. And, since the cosine terms are also nearly equal, tension force F_T must hardly change along the cable until its slope becomes quite steep. As we move from the centre of the span, where the cable is horizontal, the tension force increases only 15% where it has an upwards slope of 30 degrees. So, providing that the cable does not curve too steeply, we can consider the tension force in the cable to be nearly constant along its length.

The situation in the vertical direction is quite different. The two vertical force components $F_{T1} \sin \theta_1$ and $F_{T2} \sin \theta_2$ do not fully cancel each other. For relatively small angles (up to 20 or 30 degres), the sine of an angle varies directly in proportion to the angle. In fact, if the angle is expressed in radians (where one radian is $180/\pi$ degrees, or about 57 degrees), the sine of angle θ is equal to angle θ. The upwards component of tension forces F_{T1} is greater than the downwards ccomponent of tension force F_{T2}, giving a net upwards force which is equal to the total weight supported by the section of cable. Let's say that the load supported by the cable has a value of **w** Newtons per metre of length.

Consequently, for the net upwards force on the section of cable to be zero:

$$\underbrace{F_{T1} \sin \theta_1}_{\substack{\text{Upwards} \\ \text{force} \\ \text{component}}} - \underbrace{F_{T2} \sin \theta_2}_{\substack{\text{Downwards} \\ \text{force} \\ \text{component}}} = \underbrace{w\,\Delta L}_{\text{weight}}$$

Since the tension force is nearly constant along the cable, $F_{T1} = F_{T2} = F_T$. Furthermore, providing that the cable has a relatively shallow curvature (so that angles θ_1 and θ_2 are less than 30 degrees), then: $\sin \theta_1 = \theta_1$ and $\sin \theta_2 = \theta_2$.

So, $F_T(\sin \theta_1 - \sin \theta_2) = w\,\Delta L$

$F_T(\theta_1 - \theta_2) = w\,\Delta L$

If we define $\Delta\theta$ as $\theta_1 - \theta_2$, or the change in angle of the cable over distance ΔL, then:

Equation (1) $$F_T\,\Delta\theta = w\,\Delta L$$

Where F_T is the tension force along the cable
w is the weight per unit length supported by the cable.

If, as we have already assumed, the cable has a relatively shallow curvature, then the weight per unit length of the cable is constant. Furthermore, the weight per unit length of the roadway is also likely to be constant, and the "live load" of vehicles and pedestrians is likely to be equally distributed along the roadway. Under these conditions, which normally apply, **the angle of the cable undergoes a constant change per unit length**.

A constant change in angle with distance is characteristic of the bottom section of a circular arc (less than about 1/6 of a full circle). Imagine that the cable is a section of a circle (of radius **R**) spanning a straight line distance of length **L**, as shown here.

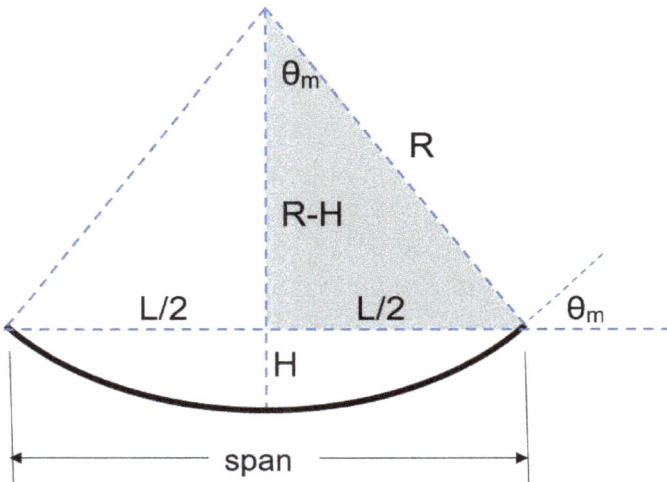

As shown in a Note in a previous chapter, we can apply the Pythagorean Theorem to the right triangle with sides of length **R** and **L** and hypotenuse (**R-H**).

Since we are considering a small section along a circlular arc, the distance **H** that the cable sags is much less than the radius **R** of the circle. In this case, we find that:

Radius of the circle, $R = \dfrac{L^2}{8\,H}$

Radius **R** can be related to the maximum angle of the cable θ_m as follows:

$\sin \theta_m = \frac{1}{2}\,L/R$

We are considering the case where angle θ_m is less than about 30 degrees, so, **$\sin \theta_m = \theta_m$**

Combining these two equations gives: $\theta_m = \dfrac{4H}{L}$

The angle θ of the cable varies uniformly with distance **x** from the centre of the arc. θ is zero at the mid-point of the cable, and reaches its maximum value θ_m at distance **L/2** from the centre. Consequently, the angle of the cable at any distance **x** from the mid-point of the arc is given by:

$$\theta = 2\left[\frac{x}{L}\right]\theta_m$$

Inserting the expression for θ_m gives: $\Delta\theta = \left[\dfrac{8\,H}{L^2}\right]\Delta L$

The angle of the cable varies directly with the distance from the mid-point of the arc. So, along

138

a short section of cable of length ΔL, the angle of the cable changes by:

Equation (2) $\quad \Delta\theta = \left[\dfrac{8\,H}{L^2} \right] \Delta L$

For a cable spanning a distance of length **2L**, and with a "sag" of **H**, we can now insert Equation (2) into Equation (1) to find the tension force in the cable:

$$\underbrace{F_T \left[\frac{8H}{L^2} \right] \Delta L}_{\Delta\theta} = w \, \Delta L$$

At the very least, the cable must be able to support its own weight. The weight per unit length of the cable, **w**, is given by its density ρ times its cross-sectional area **A** times the acceleration of gravity **g**. Inserting this into the equation above, and replacing tension force F_T with the Tensile stress P_T multiplied by the cross-sectional area of the cable, gives:

$$P_T A \left[\frac{8H}{L^2} \right] = \rho A g$$

Which tells us the tensile stress in the cable.

Equation (3) $\quad P_T = 1/8 \, \dfrac{L^2}{H} \left[\rho g \right] \qquad$ Includes only Weight of Cable

Equation (3) allows us to determine the maximum distance that can be spanned with a steel cable. As a cable approaches its maximum span, at which it is on the verge of tearing apart, the tensile stress in the cable **Pt** approaches its ultimate tensile strength P_o. Substituting P_o for **Pt** in Equation (3) and re-arranging to find the maximum value of **L**, we get:

Equation (4) $\quad$ Maximum possible span, $L = 2 \sqrt{\dfrac{2\,H\,P_o}{\rho g}}$

We can increase the maximum possible span by increasing the height **H** of the tower. But, we must keep the angle of the cable within about 30 degrees if we are to satisfy the assumptions that we have used in our calculations. Beyond this curvature, we are in a regime of rapidly diminishing returns, where a large increase in tower height yields a minimal increase in span. It turns out that the assumptions that we made are valid if the tower height **H** is no more than one-quarter the total span. Substituting this condition into Equation (4) gives:

Equation (5A) $\qquad$ Maximum possible span (cable only) $= \dfrac{P_o}{\rho g}$

If we include the weight per unit length **w** of he roadway, vehicles and pedestrians, we get:

Equation (5B) $\qquad$ Maximum possible span (cable and roadway) $= \dfrac{P_o}{\rho g + w/A}$

For a cable made of steel (with tensile strength of 1,000 MPa and density of 7,800 kilograms per cubic metre), this corresponds to a maximum possible span of 12.8 kilometres. Taking into account that the cable might need to support twice its own weight of roadway and vehicles, and assuming a safety factor of 2, gives a maximum span of around 2 kilometres. This result aligns with the 2-kilometre span of the Akashi Kaikyo Bridge, the longest suspension bridge in the world.

How might we push the boundaries to determine **the widest span that could possibly ever be achieved** (for a material of given tensile strength and density). Let's assume that we could increase the tower height without limit. The tensile stress along the cable would be a minimum at the centre of the span, and then increase as we go further out from the centre of the span (very gradually at first, and then more rapidly as the angle of the cable increaes). The increasing tensile stress could be accommodated by increasing the cross-sectional area of the cable away from the centre of the span. Then, at each point along the cable, the area of the cable would be the absolute minimum required to resist the tensile force. Each section of the cable would only be as thick as it needs to be, thereby achieving the lowest possible weight of the cable.

But this only helps up to a point. I have calculated that the required tower height (and cross-sectional area) of the cable increases very rapidly and becomes infinite as the cable span **L** approaches the value $\pi\left[\dfrac{P_0}{\rho g}\right]$.

For a steel cable, this corresponds to a theoretical maximum possible span of about 40 kilometres.

I used my calculated results to plot the shape of the cable under two scenarios:
1. Assuming that the cable has limited curvature, which gives the simplified results of Equation 3 (orange curve).

2. Assuming that the pylons are infinitely high, with the angle of the cables becoming nearly vertically at the pylons, with the cables varying in thickness to withstand higher tensile stress as the angle of the cable increases (blue curve).

As you can see, the simplified result given by Equation (3) for a cable of limited curvature is virtually identical to the result in the second scenario until the tower height exceeds one-quarter of the total span.

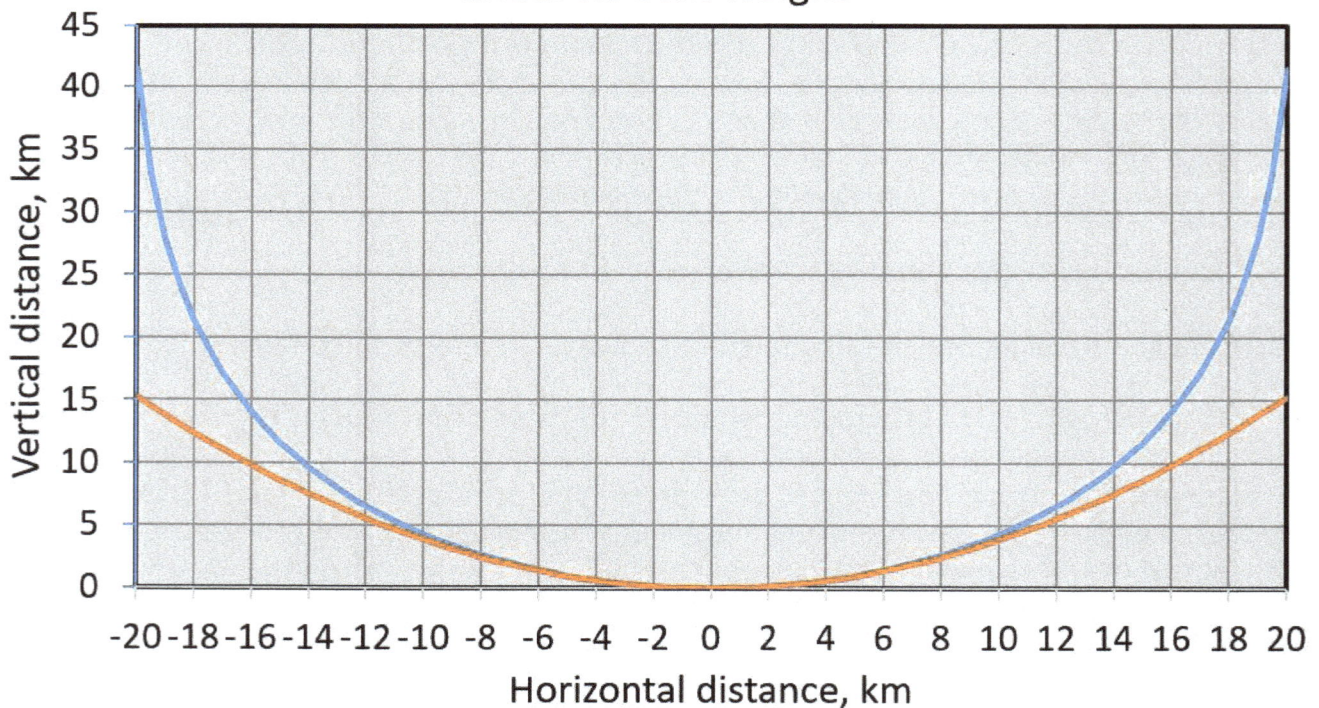

Shape of suspended cable at maximum tension under its own weight

Calculation of the maximum span of a suspended cable

The principles discussed in this chapter can be applied to calculate the maximum possible span of a bridge cable (depending upon the tensile strength and density of the material from which it is made). In fact, as I will explain, this is the maximum distance that can be spanned by any bridge.

To fully follow this derivation, you will need a basic knowledge of calculus. I have included it for readers with a solid background in mathematics, and because I have not seen it presented anywhere else. For readers with the mathematical skills, the derivation is straightforward and, I would say, elegant. The end result is extraordinarily simple and profound.

Imagine a thin flexible cable spanning a distance, starting from **-X** on one side, reaching a minimum at mid-span at **X=0**, and then reaching distance **+X**. Under the forces of its own weight and tensile forces, the cable will bend into a particular shape. The height **Y** of the cable (above its lowest point at mid-span) will be a function of distance **X**. As we did earlier in this chapter, let's consider a very small segment of the cable (of horizontal length **ΔX**), and determine the horizontal and vertical force components acting on this segment. However, instead of referring to the cable rising or falling at angle **θ**, let's now refer to the slope **m** at each particular position along the curve of the cable (for those familiar with calculus, slope **m** = dY/dX).

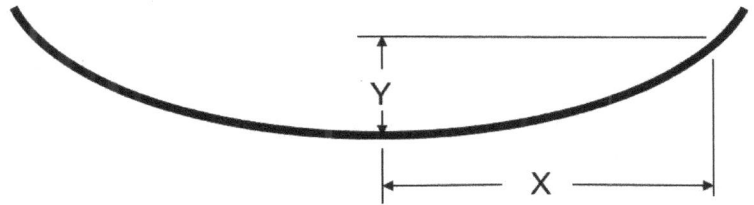

The angle **θ** of the cable is related to its slope **m** as follows:

$$\text{Cosine } \theta = \frac{1}{\sqrt{1+m^2}} \qquad \text{Sine } \theta = \frac{m}{\sqrt{1+m^2}}$$

Recall that, for the opposing horizontal force components **F₁** and **F₂** pulling on the ends of the cable segment:

$$F_1 \text{ cosine } \theta_1 = F_2 \text{ cosine } \theta_2$$

Expressing this equation in terms of the slope **m** of the cable:

Equation (A1)
$$\frac{F_1}{\sqrt{1+m_1{}^2}} = \frac{F_2}{\sqrt{1+m_2{}^2}}$$

Considering the vertical force components acting on the cable segment, we get:

$$F_1 \left[\frac{1}{\sqrt{1+m^2}} \right] \left[\frac{m_1-m_2}{\Delta X} \right] = w$$

Note that the slope **m** of the curve hardly changes (and can be regarded as constant) over a very small distance **ΔX**, but this is not true for the **change in slope m₁ – m₂**. If we consider **ΔX** to be very small (approaching zero), the change in slope **m₁ – m₂** over distance **ΔX** is given by what is called the "first derivative" dm/dX (or the "second derivative" d²Y/dX²).

The weight **w** per unit length of cable is given by its density **ρ**, area **A** and acceleration of gravity **g**. But note that, since the cable is sloping, the length of the cable segment is greater than its horizontal length **ΔX** (by a factor of 1/cosine θ).

Taking this into account, we get:

$$F \left[\frac{1}{1+m^2} \right] \frac{dm}{dX} = \rho A g$$

Now, we want to determine the **maximum possible span** of the cable. Normally, suspension cables have uniform cross-sectional area along their length, but this is sub-optimal since the tensile force in the cable increases once we move well away from the centre of the span. If we want to span the widest possible distance, we should vary the cross-sectional area of the cable so that, **at every point, the cable is just strong enough to resist the tensile forces**. In other words, the tensile stress at each point in the cable should be equal (or just a tiny bit below) the ultimate tensile strength of the material in the cable. Then:

$$\text{Cross-sectional area } \mathbf{A} = \frac{F}{Pt_o}$$

Substituting for the area of the cable in the previous equation gives:

$$\frac{1}{1+m^2} \frac{dm}{dX} = \frac{\rho g}{Pt_o}$$

We can re-arrange this equation and solve for slope **m** by using the technique of "integration". Choosing half the cable, from the centre of the span (at **X** = 0, where **m** = 0) to one end of the cable (at distance **X** from the centre):

$$\int_{m=0}^{m} \frac{dm}{1+m^2} = \int_{X=0}^{X} \frac{\rho g}{Pt_o} \, dX$$

The result is:

Equation (A2)   Slope of cable, m = tangent $\left(\dfrac{\rho g X}{Pt_o} \right)$

Where ρ is the density of the material in the cable
    g is the acceleration of gravity (9.8 metres/second^2)
    X is the horizontal distance from the centre of the span
    Pt$_o$ is the ultimate tensile strength of the material in the cable

As we would expect, the slope **m** of the cable increases as we move outwards from the centre of the span. Note that the slope becomes infinite (that is, the cable rises vertically) when the angle ($\rho gX/Pt_o$) reaches 90° – which is Π/2 when expressed in radians. Consequently, the largest possible value of distance **X** is given by the condition:

$$\frac{\rho g X}{Pt_o} = \frac{\pi}{2}$$

This means that the largest possible span (2X) is:

Equation (A3)   **Largest possible span** $= \dfrac{\pi \, Pt_o}{\rho g}$

I contend that Equation (A3) does not only set the maximum possible span for a suspension bridge, but that **Equation (A3) sets the maximum possible span for any type of bridge**.

For example, it would be impossible to build a cable-stayed bridge that exceeded this length. The roadway at the centre of the span needs to be supported by the upwards component of tension forces in cables from the top of the pylons. The weight of the cables would cause them to sag and, at the maximum span set by Equation (3A), the cables would be horizontal and unable to exert an upwards force component.

The maximum span of arch bridges would also be constrained by Equation (3A), although in this case, the span would be limited by the compressive strength of the material in the arch (rather than by its tensile strength).

Equation (A2) sets out the slope **m** of the cable in terms of distance **X** from mid-span. We can "integrate" this equation once again to find the height **Y** of the cable (relative to its lowest point at mid-span) at each distance **X** along the span. When we do this, we get:

$$\text{Equation (A4)} \qquad Y = - \frac{Pt_o}{\rho g} \left[\ln \left\{ \cosine \left(\frac{\rho g X}{Pt_o} \right) \right\} \right]$$

This equation looks complicated, but we can immediately draw some useful conclusions.

Firstly, if the spanned distance **X** is not too great (that is if $\rho gX/Pt_o$ is significantly less than one), then this complicated equation simplifies to Equation (4) derived earlier, and the cable has the shape of a parabola.

But, if we are looking to span the greatest possible distance, we want to "push the limit" and not restrict the maximum angle of the cable or height of the pylons.

Note that the height **Y** of the cable (relative to its height at mid-span) varies with the term **cosine($\rho gX/Pt_o$)**, and the height of the cable approaches infinity when this cosine term is zero. Once again, this occurs when angle ($\rho gX/Pt_o$) is 90°, or Π/2 when expressed in radians. This means that the cable rises to infinite height at the maximum span given by Equation (A3).

What would a suspension cable look like if it were spanning the largest possible distance? Near the centre, the cable would be a parabola, but further out along the cable, it begins to rise at an increasingly near-vertical slope. I plotted the shape of a cable given by Equation (A4) for cables made of typical steel (with tensile strength of 1,000 MPa and density of 7,800 kg/m³), cast iron (Pt_o = 275 MPa, ρ = 7,150 kg/m³) and glass-reinforced plastic (Pt_o = 100 MPa, ρ = 1,150 kg/m³).

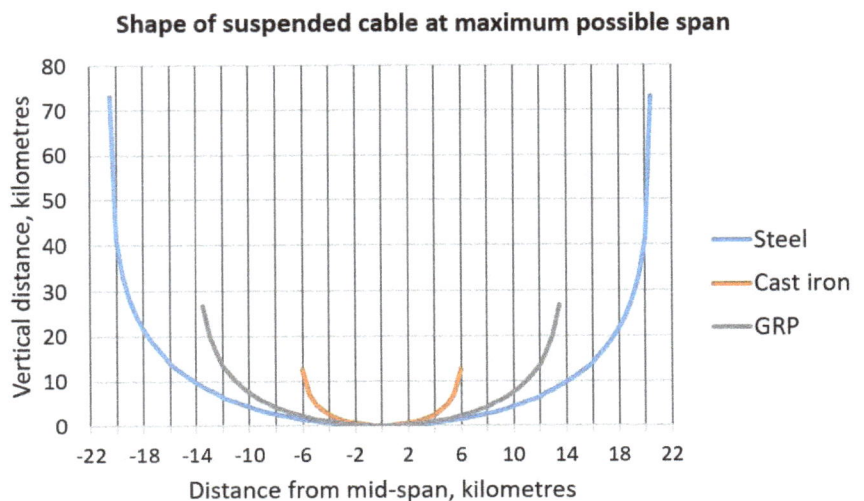

Shape of suspended cable at maximum possible span

(Vertical distance, kilometres — axis from 0 to 80; Distance from mid-span, kilometres — axis from -22 to 22. Legend: Steel, Cast iron, GRP)

The maximum span give by Equation (A3) represents an absolute theoretical maximum. It takes no account of the weight of a roadway and other structural components, vehicles crossing the bridge, the effect of high winds and a margin of safety. Obviously, a competent bridge designer would make adequate allowance for all these factors. A practical limit might be one-tenth of the theoretical maximum.

The absolute maximum span could be increased by using materials that are lighter (lower density) and stronger (higher tensile strength). High-strength, lightweight materials like some titanium alloys could be used, or others might be developed in the future, although such materials tend to be much more expensive than steel and more difficult to fabricate. Thus, while it may be possible to build bridges with longer spans, it may not be economically feasible.

Further into the future, mankind might inhabit other planets or moons in our solar system, and the ability to build structures spanning large distances will depend on the gravitational attraction at the surface. The acceleration of gravity **g** on the moon is only 16% that on Earth, and for Mars, **g** is just over a third as much as on Earth. Weaker gravity would make it much easier to build large structures that support their own weight. However, this would be offset by much greater difficulty and expense of transporting materials from the Earth, or producing them locally, as well as the many other challenges of working and living beyond the protective environment of Earth.

Martin Gellender has always been fascinated by science. He grew up in New York City during the period after World War II, living in an environment that was economically-poor but intellectually-enriched. This was a time of awe, when people marveled at the development of antibiotics, synthetic fabrics, skyscrapers, nuclear energy, rocket propulsion, jet aircraft and plastics. The world seemed to be on the verge of a scientific and technological revolution - and indeed, it was.

The high-rise apartment block where he lived in Lower Manhattan was basically a run-down slum, but its location was about as close as you could get to . . . the centre of the modern world. Everything was within walking distance – China Town, Little Italy, Madison Avenue, the United Nations, Greenwich Village, Fifth Avenue, Central Park, you name it.

Weekends with his family were spent on long walks to all of these places, and more. These walks often included riding a ferry across New York harbour, walking across the Williamsburg Bridge to Brooklyn, visiting the Hayden Planetarium or Museum of Natural History, Gilbert Hall of Science, browsing through the largest bookshop in the world (Barnes and Noble), touring US Navy ships when they berthed in New York, going to the observation deck of the Empire State Building (then, the tallest building in the world) or browsing through numerous shops selling "army surplus equipment" (everything from generators to bomb sights) and electronic components. These all-day walks often finished with a visit to his uncle and aunt's house. There, discussions with his Uncle Sol, father and older brother often revolved around scientific developments that were underway and the possibilities for the future. Being the youngest, Marty struggled to understand what was said, but that only added to his curiousity and sense of wonder.

In 1957, the Soviet Union launched the Sputnick Satellite, and Americans were stunned that they had fallen behind their technological rival. They responded by pouring resources into scientific education. Five years later, when Marty attended a specialised high school ("Brooklyn Tech"), the facilities (machine shops, a foundry, chemistry laboratories) were state-of-the-art then (and probably better than what you would find now in any high school in Australia).

Marty went on to graduate with a bachelors degree in Chemistry, worked for a pharmaceutical company, and did a PhD (City University of New York). He moved to Canada, where he worked as a science writer for a chemical engineering magazine (and married an Aussie), and then moved to England to start a new chemistry magazine. In 1982, he relocated to Brisbane and spent most of his career in the Queensland Government. He played an instrumental role in negotiating an agreement between the Queensland Government and CSIRO to establish the Queensland Centre for Advanced Technology (QCAT) in Brisbane, and in setting up an Energy Information Centre. Most recently, he managed a grants program that funded companies developing energy-efficient and water-saving technology.

Marty is still trying to understand how the world works, and takes great pleasure sharing his knowledge and passion with others through the University of the Third Age, and in writing this book.

www.ingramcontent.com/pod-product-compliance
Lightning Source LLC
Chambersburg PA
CBHW041154220326
41598CB00045B/7424